Mathematics
at the Meridian

Numerical Analysis and Scientific Computing Series

Series Editors:
Frederic Magoules, Choi-Hong Lai

About the Series

This series, comprising of a diverse collection of textbooks, references, and handbooks, brings together a wide range of topics across numerical analysis and scientific computing. The books contained in this series will appeal to an academic audience, both in mathematics and computer science, and naturally find applications in engineering and the physical sciences.

Iterative Splitting Methods for Differential Equations
Juergen Geiser

Handbook of Sinc Numerical Methods
Frank Stenger

Computational Methods for Numerical Analysis with R
James P Howard, II

Numerical Techniques for Direct and Large-Eddy Simulations
Xi Jiang, Choi-Hong Lai

Decomposition Methods for Differential Equations
Theory and Applications
Juergen Geiser

Mathematical Objects in C++
Computational Tools in A Unified Object-Oriented Approach
Yair Shapira

Computational Fluid Dynamics
Frederic Magoules

Mathematics at the Meridian
The History of Mathematics at Greenwich
Raymond Flood, Tony Mann, Mary Croarken

For more information about this series please visit: https://www.crcpress.com/Chapman--HallCRC-Numerical-Analysis-and-Scientific-Computing-Series/book-series/CHNUANSCCOM

Mathematics
at the Meridian
The History of Mathematics at
Greenwich

Edited by
Raymond Flood, Tony Mann
and Mary Croarken

CRC Press
Taylor & Francis Group
Boca Raton London New York

CRC Press is an imprint of the
Taylor & Francis Group, an **informa** business
A CHAPMAN & HALL BOOK

CRC Press
Taylor & Francis Group
6000 Broken Sound Parkway NW, Suite 300
Boca Raton, FL 33487-2742

International Standard Book Number-13: 978-0-8153-6879-3 (Hardback)
International Standard Book Number-13: 978-0-367-36272-0 (Paperback)

Library of Congress Control Number: 2019948013

Visit the Taylor & Francis Web site at
http://www.taylorandfrancis.com

and the CRC Press Web site at
http://www.crcpress.com

Contents

Foreword

I am delighted for several reasons that this account of the history of mathematics at Greenwich is appearing in the *Chapman & Hall CRC Numerical Mathematics and Scientific Computing* series.

As this book shows, numerical mathematics and computation have been a focus of mathematicians working at Greenwich since the foundation of the Royal Observatory in 1675. John Flamsteed and his successors gathered data and developed mathematical understanding with a view to enabling mariners to compute their position at sea. Nevil Maskelyne's pioneering work on computation for the Nautical Almanac anticipated concerns of those doing numerical mathematics and scientific computing in the twentieth and twenty-first centuries. George Biddell Airy played a significant part in the story of Charles Babbage's calculating machines. Leslie Comrie (from whom the present-day computing and mathematics research seminar series at the University of Greenwich takes its name) was a major pioneer in using machines to carry out mathematical computation in the second quarter of the twentieth century. Today, numerical mathematical modelling is a major area of interest for mathematicians at the University of Greenwich.

The reader will learn about contributions to numerical mathematics and computation made at Greenwich over more than a third of a millennium. They will also discover other significant contributions made by Greenwich mathematicians in various areas of mathematics, mathematics education, and public engagement.

I hope that the story of mathematics at Greenwich will inspire readers and encourage visitors to explore the rich mathematical heritage of Greenwich by visiting the Royal Observatory, the National Maritime Museum, the Painted Hall of the Old Royal Naval College, and the historic *Cutty Sark*, now preserved and proudly displayed with mathematical contributions from my colleagues.

Choi-Hong Lai
Professor of Numerical Mathematics, University of Greenwich
August 2019

Preface

We owe thanks to many people. We are grateful to our contributors for their chapters and all their help in completing this project. At the University of Greenwich, Liz Bacon, Chris Bailey, Judith Burnett, Choi-Hong Lai, Mary McAlinden, Robert Mayor, Kevin Parrott, and Mark O'Thomas encouraged the project, Phil Hudson assisted with preparing the images, Irene Barranco Garcia and Karen Richardson gave advice, Neil Saunders helped with editing, and many other colleagues provided support. Elsewhere, June Barrow-Green, Alan Camina, Alex Craik, Geoffrey Howson, the late Clive Kilmister, Peter Neumann, Susan Oakes, and Jane Secker provided information about mathematics at the Royal Naval College and other advice. Rebecca Young, archivist at the Royal Hospital School, Elizabeth Fisher at the London Mathematical Society and Mike Price, librarian for the Mathematical Association, assisted by finding photographs, and Vanilla Beer and Diana Pooley kindly gave permission for the reproduction of their drawings.

We are very sorry that Bernard de Neumann did not live to see the publication of this book. Bernard had great enthusiasm for the history of the Royal Hospital School and took delight in unearthing new information about its mathematicians, and we are very pleased that despite his poor health he was able to complete his contribution to the book.

Finally, Sarfraz Khan and Callum Fraser at CRC Press have been supportive, prompt with advice and information, and generous in their understanding of our delays.

RGF, TM and MGC
January 2019

Editors

Raymond Flood has spent most of his academic life promoting mathematics and computing to adult audiences, mainly through his position as University Lecturer at Oxford University, in the Continuing Education Department and at Kellogg College. In parallel, he has worked extensively on the history of mathematics, producing many books and other educational material.

He is Emeritus Fellow of Kellogg College, Oxford, having been Vice-President of the College and President of the British Society for the History of Mathematics before retiring in 2010. He was Gresham College Professor of Geometry from 2012 to 2016. He is a graduate of Queen's University, Belfast; Linacre College, Oxford; and University College, Dublin, where he obtained his PhD.

Tony Mann is Director of Greenwich Maths Centre at the University of Greenwich, where he has taught for many years. He is a past President of the British Society for the History of Mathematics, and has published on mathematics in Victorian Scotland and on mathematics in fiction. He is a former Visiting Professor of Computing Mathematics at Gresham College.

As a former undergraduate at Pembroke College, Cambridge, with which Burnside was closely connected, Tony is pleased to have edited, with Peter M. Neumann and Julia Tompson, *The Collected Papers of William Burnside* (Oxford University Press, 2004).

Mary Croarken started her academic career at the University of Warwick with a degree in Computer Science and then a PhD in History of Computing. Mary has subsequently held a series of academic fellowships including the Sackler Fellowship at the Centre for Maritime Research at the National Maritime Museum (NMM) 2000–2002. While at NMM, she helped to organise a joint NMM and British Society for the History of Mathematics conference called *Greenwich: Some Mathematical Connections*, which provided the inspiration for this book. Mary continues her academic interests in human computers, but now works full time supporting health research in the National Health Service (NHS).

Contributors

Noel-Ann Bradshaw undertook her mathematics degree as a mature student before obtaining a lectureship in mathematics and operational research at the University of Greenwich and taking on roles such as Faculty Director of Employability. She holds a Professional Doctorate on Innovations within Higher Education Mathematics Education. She is now Vice-President (Communications) for the Institute of Mathematics and its Applications and previously served on the Council for the British Society for the History of Mathematics for seven years, first as Meetings Secretary and then as Treasurer. A former editor of the mathematics education journal *MSOR Connections*, she was keynote speaker at the CETL-MSOR conference on mathematics education in 2018 and at the IMA's undergraduate conference Tomorrow's Mathematicians Today in 2019; the latter is a conference she herself initiated in 2010. As well as workshops for school students on topics in mathematics, operational research, and the history of mathematics (sometimes in historical costume), her public engagement activities have included academic talks on Florence Nightingale's statistical diagrams and her use of data to influence policy. Her invited chapter on Nightingale appears in *Women in Mathematics* (ed. by Beery et al, 2018). Inspired by the lives that Nightingale changed through the effective use and communication of data, she is now working as a data scientist for Sainsbury's Argos.

Allan Chapman is a historian of science at Oxford University, specialising in astronomy and medicine. He obtained a DPhil at Oxford and has been awarded honorary doctorates by the University of Central Lancashire and Salford University, and an Outstanding Alumnus Award by Lancaster University. In 2015 he was presented with the Jackson-Gwilt Medal by the Royal Astronomical Society. He lectures widely, has made two TV series, and has written a dozen books and numerous academic papers. His books include *Dividing the Circle* (1990, 1995), *Astronomical Instruments and Their Users* (1996), *The Victorian Amateur Astronomer* (1996, 2017), *England's Leonardo* [Robert Hooke] (2005), and *Stargazers* (2014). He is fascinated by tools and technologies, and has made working replicas of many early astronomical instruments.

Bernard de Neumann was educated at the Royal Hospital School and then Birmingham University. Following his MSc, he joined Marconi Research Laboratories at Great Baddow, Essex, where he spent almost 25 years working as a 'pure' applied mathematician. His work included modelling Concorde, in order to show that it could land safely, and modelling various missile systems. While there, he became interested in the mathematics of reliable systems and the optimal logistics of supporting military and large civil operational systems.

In 1986 he was made an Honorary Visiting Professor in the Mathematics Department at City University, London, later joining the department as a Professor in 1988. In 1989 he was appointed Research Professor at City in the Centre for Software Reliability.

He left City in 1995 to join the Ministry of Defence. He retired in 1999 becoming, for a while, a non-executive director of the Essex Health Authority. He then devoted much of his time to researching naval history. Bernard died in 2018.

Richard Dunn is Keeper of Technologies and Engineering at the Science Museum. Until 2019 he was Senior Curator for the History of Science at Royal Museums Greenwich, where his research focused on the history of marine navigation. His publications include *The Telescope: A Short History* (2009), *Finding Longitude: How Ships, Clocks and Stars Helped Solve the Longitude Problem* (with Rebekah Higgitt, 2014), *Navigational Enterprises in Europe and its Empires, 1730–1850* (ed. with Rebekah Higgitt, 2015) and *Navigational Instruments* (2016).

J. Helen Gardner is a retired teacher, having taught mathematics, English, and history at all levels. She has a keen interest in the history of mathematics, pursuing a higher degree with the Open University on 'Using the History of Mathematics in Primary Education'. A former Membership Secretary for the British Society for the History of Mathematics, she has promoted the use of the history of mathematics as a vehicle to help students' understanding of the subject, both in articles and in lectures at teacher development conferences. She has co-authored six articles on Thomas Archer Hirst, as well as given many talks about him, both in the UK and abroad.

Dr Peter M. Neumann is a retired university mathematician. He was born in 1940 in Oxford, but his school years were spent in Hull, East Yorkshire. In 1959 he returned to Oxford to read mathematics and has been a member of The Queen's College and of Oxford University ever since.

He holds BA, MA, DPhil and DSc degrees earned (mostly) from Oxford, and an honorary DSc degree from the University of Hull. He has collected a few prizes from time to time for his research, his writing and his teaching. His research has been in algebra (mostly the theory of groups) and its history. He has written about 100 articles and a few books. As a teacher, he covered most of the undergraduate pure mathematics syllabus – anything of interest to himself and to students – and supervised 40 doctoral students. Until 28th February 2018, when he suffered a stroke, he continued to teach in his college and university. He also offered mathematical enrichment sessions to gifted schoolchildren aged between 10 and 17. He was appointed OBE for services to education in 2008.

He contributed to the foundation of the UK Mathematics Trust and served from 1996 to 2004 as its first chairman. In his time, he has also served as Vice-Chairman of the London Mathematical Society, President of the British Society for the History of Mathematics, and President of the Mathematical Association.

Doron D. Swade (MBE) is an engineer, historian, museum professional, and a leading authority on the life and work of Charles Babbage, nineteenth-century computer pioneer. He has studied physics, electronics engineering, philosophy of science, machine intelligence and history at various universities including Cambridge University and University College London. He was Senior Curator of Computing at the Science Museum, London, and later Assistant Director & Head of Collections.

He has published three books and many articles on curatorship, museology and history of computing. He is currently a Research Fellow (Computer Science) at Royal Holloway, University of London.

Robin Wilson is an Emeritus Professor of Pure Mathematics at the Open University, and of Geometry at Gresham College, London. A former Fellow of Keble College, Oxford University, he is currently a Visiting Professor at the London School of Economics. A former President of the British Society for the History of Mathematics, he has written and edited about 40 books on the history of mathematics and on graph theory and combinatorics. Involved with the popularization and communication of mathematics and its history, he has been awarded two prizes from the Mathematical Association of America for his 'outstanding expository writing', and he was awarded the 2017 Stanton Medal by the Institute of Combinatorics and its Applications for outreach activities in combinatorial mathematics.

0

Introduction

Raymond Flood and Tony Mann

CONTENTS

0.1 Introduction

In 1997 UNESCO declared Greenwich in south-east London a World Heritage Site in recognition of its beautiful buildings and its importance in world cultural and scientific history. Greenwich joined a list of over 850 such locations across the world, ranging from the Great Wall of China to the Taj Mahal and from the historic centre of Florence to the Acropolis.

At Greenwich we find what is often described as the 'centre of space and time'. There you can stand astride the prime meridian with one foot in the western hemisphere and the other in the eastern hemisphere as well as checking your watch against Greenwich Mean Time.

The main components of this world heritage site are

- The National Maritime Museum (NMM) – dedicated to the illustration of Britain's long sea-faring tradition.
- The Queen's House – designed by Inigo Jones (1573–1652), designed on Palladian geometrical principles for Anne of Denmark and completed for Henrietta Maria, and now mainly a portrait gallery.
- The Old Royal Naval College, originally Greenwich Hospital – now a base of the University of Greenwich and Trinity Laban Conservatoire of Music and Dance.

- The Royal Observatory Greenwich – founded in 1675 and crucial in the quest for an effective means of finding longitude at sea.
- Greenwich Park – London's oldest enclosed Park.
- The *Cutty Sark* – the fastest sailing ship of her day.

An anonymous painting of Greenwich (after 1750), held by Tate, is reproduced on the cover of this book. There are also two paintings of Greenwich by Canaletto (Giovanni Antonio Canal) (1697–1768): *A view of Greenwich from the River* (c. 1750–1752), currently on loan to Tate,[1] and *Greenwich Hospital from the north bank of the Thames* (c. 1750–1752), held by Royal Museums Greenwich and currently displayed in the Queen's House.[2]

Greenwich is a popular destination for tourists visiting the capital: it can be reached by boat down the Thames from central London and its historic buildings, royal park, market, and other attractions have much to offer visitors.

While the Royal Observatory, the Prime Meridian, and the use of Greenwich Mean Time are reminders that Greenwich has a rich astronomical history, many visitors may be less aware of the importance of Greenwich in the development of numerical mathematics. But from the foundation of the Observatory in 1675 computation has been central to the mathematics done at Greenwich. Early Astronomers Royal worked to gather and improve the astronomical data necessary to improve navigation. In the eighteenth century Nevil Maskelyne created algorithms and error-correcting mechanisms for the human computers who calculated the navigational tables for the *Nautical Almanac*. Astronomer Royal George Biddell Airy was a key player in the nineteenth-century debate about Charles Babbage's calculating engines. In the second quarter of the twentieth century Leslie Comrie mechanized the production of navigational tables using mechanical calculating machines. And in the twenty-first century pioneering work in numerical mathematics continues at the University of Greenwich, including mathematical modelling which has helped preserve the maritime icon, the *Cutty Sark*.

In this book we explore how Greenwich has hosted fundamental and influential investigations in the mathematical sciences from the late seventeenth century to the present day. While numerical mathematics plays the leading part, pioneering pure mathematicians worked at the Royal Naval College, and many Greenwich mathematicians made contributions to mathematics education and public engagement, so these also feature in our story of mathematics at Greenwich.

0.2 Astronomers Royal and Finding Longitude

It started with a practical computational problem of the utmost importance. As Mary Croarken recounts in Chapter 2, when King Charles II established the post of Astronomer Royal in 1675, the terms of reference included the instructions that the Astronomer Royal was: 'to apply yourself with the most exact Care and Diligence to the rectifying of the Tables of the Motions of the Heavens, and the Places of the fixed Stars, so as to find out the so much desired Longitude at Sea, for the perfecting of the Art of Navigation'.

In order to enable ships at sea to determine their longitude it would be sufficient for them to be able to compare their 'local' time with the time at Greenwich. This could be achieved either by carrying an accurate chronometer or, computationally, by using the position of the moon against the fixed stars. Neither of these would be easy since it would be necessary to improve the accuracy of the tables describing the moon's motion relative to the fixed stars to new levels of accuracy or alternatively to improve dramatically both the accuracy and portability of time keeping. On land at least the time-keeping problem was solved, by 1670, either astronomically or with the further development of pendulum clocks by Robert Hooke and William Clement, but these clocks could not keep good time at sea. It was these arguments, proposed by John Flamsteed, that persuaded the King to found the Royal Observatory with Flamsteed as first holder of the office that was to become internationally known as Astronomer Royal (Figure 0.1).

FIGURE 0.1 John Flamsteed from Flamsteed's *Historia Coelestis*. (Reproduced from E. Walter Maunder, *The Royal Observatory Greenwich: A Glance at Its History and Work* (London: The Religious Tract Society, 1900).)

High-quality data is essential when numerical computation is used, and Flamsteed concentrated on observations and data-gathering rather than on theoretical investigations in order to answer the outstanding questions of his day – the truth or otherwise of the Copernican theory, the motion of the moon, and the distances of the stars from Earth. Crucial to his approach was advancement in instrumentation to facilitate more and more accurate observations, and in this approach he saw himself in the tradition of Tycho Brahe and Johannes Hevelius. However unlike both of these predecessors, Flamsteed incorporated the telescope into his astronomical measuring instruments. As Allan Chapman discusses in Chapter 1 there is also an English dimension to the influences on Flamsteed and the approaches that he took. This centred around Jeremiah Horrocks, William Gascoigne and William Crabtree, and the Townley family based in Lancashire and Yorkshire, and Flamsteed enthusiastically embraced his countrymen's practical advances. Indeed Chapman sees Flamsteed's practical approach as beginning a tradition which has persisted down to the twentieth century.

The beautiful Royal Observatory was built by Sir Christopher Wren, coming in at less than 5% over the original budget of £500. It has a superb situation, 6 miles east of London, overlooking the Thames and towards the Essex marshes and, of course, at the time of its foundation in 1675, would have been spared today's light polluted skies. (Figure 0.2).

Flamsteed, with the support of his friend Sir Jonas Moore, Master of the King's ordnance, set about equipping the Observatory with the central aim of increasing angle-measuring ability to a greater level of accuracy so as to enable him to record the moon's complicated orbit among the fixed stars. Sir Jonas provided money to help support the equipping of the observatory as did Flamsteed himself, who claimed to have spent £2,000 pounds in buying equipment – a considerable sum equivalent to 20 years of his salary. Chapter 1 describes the instruments and how they were used to obtain precise and reliable measurements of the moon, stars, and planets through

FIGURE 0.2   The Royal Observatory. (Reproduced from E. Walter Maunder, *The Royal Observatory Greenwich: A Glance at Its History and Work* (London: The Religious Tract Society, 1900).)

FIGURE 0.3   The 'Camera Stellata' or Octagon Room of the Observatory. (Reproduced from E. Walter Maunder, *The Royal Observatory Greenwich: A Glance at Its History and Work* (London: The Religious Tract Society, 1900).)

a process of continuous observing and re-observing using different instruments and techniques (Figure 0.3).

That Flamsteed's observations were well regarded is illustrated by the efforts, starting in 1704, that Sir Isaac Newton went to in order to get hold of them. This eventually led to an uncorrected version of his Greenwich observations being published by Edmond Halley in 1712. Flamsteed bitterly resented this enforced publication and eventually was to burn as many copies of the work as he could lay his hands on. He then in 1715 decided to publish his life's work at his own expense, employing reliable men to recalculate the data and independently check the proofs from the printers. However, on New Year's Eve 1719, he died before the project could be completed, and it was his widow Margaret who, with determination and energy, saw Flamsteed's work eventually published.

Margaret's determination to publish her husband's life work might have been reinforced by the appointment of Edmond Halley as Flamsteed's successor in 1720 – an appointment she could be forgiven for viewing as insulting to her husband's memory.

Halley's ambition was to make practical the lunar distances method of calculating a ship's position at sea, by gathering improved data on the motion of the moon. His first task was to re-equip the Observatory, since Flamsteed's executors had removed all his instruments. Halley persuaded the Government to provide the Observatory's instruments, and as a result they remained in the ownership of the Observatory and many can be seen today in the NMM. He was however still only paid a salary of £100 a year, and it was not until the appointment of George Airy over a hundred years later that the Astronomer Royal was paid a proper professional salary.

Even though he was 64 when appointed Halley was an energetic and highly respected astronomer who also held the Savilian Chair of Geometry at Oxford. Chapter 1 describes how he used a transit, quadrant, and regulator clock manufactured by the craftsman-scientist, George Graham, to make his observations. However, in spite of these efforts, Halley was unable to observe the moon with sufficient accuracy to determine longitude and hence win the £20,000 prize which had been established by the Longitude Act of 1714.

FIGURE 0.4   James Bradley. (Reproduced from E. Walter Maunder, *The Royal Observatory Greenwich: A Glance at Its History and Work* (London: The Religious Tract Society, 1900).)

Halley's successor as Astronomer Royal on his death in 1742 was his Oxford colleague, James Bradley, Savilian Professor of Astronomy (Figure 0.4). Bradley was already renowned for his discovery, published in 1728, of the aberration of light noticed when observing the star Gamma Draconis and as a result demonstrating the motion of the earth about the sun two centuries after Copernicus had suggested it in his *De revolutionibus orbium coelestium* of 1543. To achieve this Bradley set new standards in the precision of his observations. This precision brought another remarkable discovery in celestial mechanics in 1748 – the Nutation, a significant 'nodding' of the earth's orbital system with the moon due to the 18-year lunar cycle around the earth. At this time Allan Chapman argues the Royal Observatory was the foremost observatory in Europe to have a productive and effective collaboration between scientists and craftsmen-scientists in their investigation of how materials responded to changes in temperature or other atmospheric factors with the aim of manufacturing ever more precise instruments. The profession of scientific instrument-maker – for profession it was – was well regarded, socially accepted and financially rewarding with practitioners such as Thomas Tompion, George Graham, John Bird, and Jesse Ramsden having close association with the Royal Observatory. It was John Bird who re-equipped the Observatory for Bradley with an approach that tried to eliminate or compensate for all sources of error either in the differential expansion of the parts of an instrument made from different materials when the temperature varied or in the variation caused by atmospheric fluctuations. Their approach also had cross-checks between the various instruments and clocks. Bradley's methods were adopted as standard practice in all observatories. The result was to achieve an accuracy of plus or minus 1 arc-second in the meridian and an international reputation for accuracy as acknowledged by, for example, Frederick Wilhelm Bessel whose view was that Bradley's observations were the earliest that could be relied upon for investigating long-term gravitational effects.

The motivation for establishing the Royal Observatory was to enable the finding of longitude through astronomical observations and in particular by observing the motion of the moon. It took over 80 years from its foundation for this to be achieved, but it was Tobias Meyer, from Gottingen observatory, who produced the first usable set of lunar tables, although when his results were compared with Bradley's, there was only a difference of 2 arc-seconds. Five years after Bradley's death in 1762 his successor but one Nevil Maskelyne produced, on the basis of Greenwich and Mayer's observations, the first *Nautical Almanac* – a set of lunar tables that navigators could use to fix their position – just over a century after the Observatory was founded. The set of tables only cost a few shillings and was therefore much, much cheaper than the alternative method using a Harrison Chronometer which cost over £100. In the *Nautical Almanac* Maskelyne pre-computed lunar distances, thus substantially reducing the amount of calculation needed to determine longitude after observations had been taken on board ship. Mary Croarken, in Chapter 2, reviews Maskelyne's involvement in the award of the Longitude prize to Harrison and Meyer and also tells the fascinating story of how Maskelyne set up a network of human computers, geographically distributed and communicating by post, to calculate the tables according to procedures devised by Maskelyne. The human computers were of three types, a 'computer' and an 'anti-computer' who worked independently on the same calculation and then sent their calculations to a 'comparer' who did what the title suggests, following up on and eliminating discrepancies. The system was dedicated to obtaining reliable accurate tables, and it was successful in doing this – even on one occasion detecting collusion between a computer and anti-computer! The computers were mainly men and had other occupations, for example as clerks, vicars, surveyors, or schoolmasters.

Measurements of the position of the moon relative to a given star had, however, to be made on board a ship which was rarely a steady base for observing. It was Jesse Ramsden and then Edward Troughton who developed a cheap, accurate, and efficient manufacturing process for the production of machine graduated sextants which could be used at sea.

Chapter 2 also traces the influence of the *Nautical Almanac* in the establishment of Greenwich as the international meridian for longitude zero in 1871, followed in 1884 by the adoption of Greenwich as the international prime meridian – the 'centre of time'.

The fourth Astronomer Royal Nathaniel Bliss, coming between Bradley and Maskelyne (Table 0.1), died after 2 years in office, having been appointed at the age of 62. His only significant achievement as Astronomer Royal was to observe an annular eclipse of the sun.

TABLE 0.1

Names and Dates of the Astronomers Royal

- 1675–1719 John Flamsteed
- 1720–1742 Edmond Halley
- 1742–1762 James Bradley
- 1762–1764 Nathaniel Bliss
- 1765–1811 Nevil Maskelyne
- 1811–1835 John Pond
- 1835–1881 George Airy
- 1881–1910 William Christie
- 1910–1933 Frank Dyson
- 1933–1955 Harold Spencer Jones
- 1956–1971 Richard van der Riet Woolley
- 1972–1982 Martin Ryle
- 1982–1990 Francis Graham-Smith
- 1991–1995 Arnold Wolfendale
- 1995–*present* Martin Rees, Baron Rees of Ludlow

0.3 Astronomers Royal after the Longitude Prize

The sixth Astronomer Royal, John Pond, held office for nearly a quarter of a century before his resignation in 1835. He was a well-regarded practical astronomer with contributions particularly to declination measurements. During his term of office the number of assistants at the Observatory grew from one to six. He was buried in the same tomb at Lee in Kent as his predecessor Edmond Halley (Figure 0.5).[3]

If Pond is little remembered, the same cannot be said of George Biddell Airy who, as Doron Swade discusses in Chapter 3, suffered a bad press over his failure to discover the planet Neptune. Airy had a brilliant undergraduate career at Trinity College, Cambridge, and in 1826 was appointed Lucasian professor of mathematics and then in 1828 Plumian professor of astronomy and Director of the newly created Cambridge University Observatory. Airy was not independently wealthy and was dependent on his salary for his living. In 1835 he obtained the post of Astronomer Royal and negotiated a salary of £800 per annum. He was by nature very methodical, organized, scrupulous in his record-keeping, and strict on public accountability. For him the role of the Royal Observatory was the practical one of observation to aid navigation and astronomy. This and his other work together with family concerns may help explain why Airy did not engage with John Couch Adams' calculations on the irregularities of the orbit of Uranus and the hypothesis that they were caused by a new planet. So it was left to Johann Galle in Berlin, working from calculations made by Urbain Leverrier, to discover Neptune.

Doron Swade takes a fresh look at Airy and at the impact that he had on Victorian science in his role essentially as the government's chief scientific adviser – in particular on the development of calculating engines. Airy was very sceptical about the usefulness of mechanical computing engines and played an important role in the Government's decision to end support for Charles Babbage's projects: Babbage's engines, uncompleted in his lifetime, are now seen as forerunners of the modern computer. Doron Swade investigates the background to the interaction between Babbage

FIGURE 0.5 John Pond. (Reproduced from E. Walter Maunder, *The Royal Observatory Greenwich: A Glance at Its History and Work* (London: The Religious Tract Society, 1900).)

and Airy and identifies the reasons behind Airy's lack of enthusiasm. Although Babbage had originally conceived of the engines for their use in mathematical investigations, he and his supporters promoted their potential benefit in producing error-free mathematical tables. And it was essentially these supposed utilitarian grounds to which Airy objected, asserting that such machines would be useless in compiling any part of the Nautical Almanac. He felt that existing methods were sufficiently accurate, that there was little demand for new tables, and that there was insufficient economic benefit to justify the high capital cost of the new machines. Furthermore, he argued that such machines could only perform part of the task as it was still necessary for experienced mathematicians to develop the analysis and also that such devices would be too complicated to be used by the clerks involved in the computations. Finally he criticized the motives of those developing such machines as not being concerned with the true needs of (human) computers but being driven by their own mathematical and technology interests. These arguments all had force and were very influential. Doron Swade unpacks these issues and concludes with a thoughtful re-evaluation of Airy.

In Chapter 4 Tony Mann looks at the subsequent history of the Greenwich Observatory. Airy's successor was William Christie, and it was during his tenure that the International Meridian conference in Washington DC voted to make Greenwich the prime meridian, a development partly due to his fostering of international collaboration. He also introduced photography as a new method of mapping the sky as well as spectroscopy as a new observational technique.

Frank Dyson succeeded Christie in 1910 moving from the position of Astronomer Royal for Scotland and was instrumental in the observation of the total solar eclipse of the sun on 29 May 1919 with a view to testing Einstein's general theory of relativity. His legacy can still be heard today as he introduced the 'Greenwich time signal' – the six familiar 'pips' with the final one marking the hour. As Tony Mann comments, this is another icon of Greenwich's contribution to navigation. Dyson remained in post until 1933 and interacted, sometimes not always harmoniously, with other astronomers, and it was also during his period that industrialization began to affect the observational work of the Royal Observatory, with the electrification of the local railway resulting in the magnetic investigations being moved to Abinger in Surrey. However, it was the tenth Astronomer Royal, Harold Spencer Jones who had to confront the deteriorating observing conditions at Greenwich due to light pollution, and the decision was made to move the Observatory to a new home at Herstmonceux Castle in Sussex. The move was delayed by the Second World War and in fact was not completed until 1956 at the start of Richard van der Riet Woolley's term of office. The site at Greenwich then passed to the NMM. Thirty years later in 1986 the Royal Greenwich Observatory made its final move from Herstmonceux to Cambridge and was closed, after a 323-year history, in 1998 with its staff and functions being distributed between Cambridge, Edinburgh, the Rutherford Appleton Laboratory in Didcot, and the NMM. The site at Greenwich is now again known as the Royal Observatory, Greenwich.

0.4 The Royal Hospital School

In Chapter 5 Bernard de Neumann examines the mathematics training given from 1712 to potential seamen at Greenwich at the Royal Hospital School, originally in part of the then Greenwich Hospital (now the Old Royal Naval College) and subsequently in what is now the NMM. The School moved to Holbrook in Suffolk in 1933.

Bernard de Neumann discusses the role of mathematics in naval education and tells us about some of the notable teachers of mathematics, in particular Edward and John Riddle, successive headmasters of the Royal Hospital School.

0.5 The Royal Naval College

The buildings occupied by the Royal Naval College – by the Thames and at the bottom of the hill from the Observatory – were planned from 1694 and, as discussed by Bernard de Neumann in Chapter 5, were originally built for the care of seamen and their dependents and for the education of the children of seamen. This was the beginning of the teaching of mathematics at Greenwich. Sir Christopher Wren was the architect, and the design was completed by architects including Nicholas Hawksmoor, Sir John Vanbrugh, and Thomas Ripley. The infirmary designed in the 1760s by James 'Athenian' Stuart continued to function as Dreadnought Seaman's Hospital after the closure of the Royal Hospital until 1986 (Figure 0.6).

In the nineteenth century the numbers needing care decreased – from a peak of 3,000 in 1813 – and Greenwich Hospital closed for this purpose in 1869. Its new role

FIGURE 0.6 Overhead view of the Old Royal Naval College. (Courtesy of the University of Greenwich.)

was as the Royal Naval College, responsible for the training of naval officers and of selected naval architects and marine engineers, and established by order in Council in 1873. The aim of the curriculum was to equip its students with the theoretical and scientific underpinning for their profession of naval officer. The move of the Royal Naval College to Greenwich owed much to the wish of the Greenwich MP, William Ewart Gladstone, to provide jobs in his constituency in order to save his parliamentary seat.

Tony Mann's Chapter 6 considers the mathematicians and mathematics teaching at the Royal Naval College (Table 0.2) and its involvement in the public communication of mathematics. (The first Director, Thomas Archer Hirst, and the pioneering pure mathematician William Burnside are discussed in separate chapters.) Chapter 6 examines the appointments to the mathematical positions, looking at the contributions of the staff employed by the College and imparting a vibrant impression of life at the College from both the student and staff perspective.

TABLE 0.2

Professors of Mathematics at the Royal Naval College

Robert Kalley Miller	1873–1889
Carlton John Lambert	1877–1909
William Burnside	1885–1919
Charles Godfrey	1920–1924
George Arthur Witherington	1924–1934
Louis Milne-Thomson	1934–1956
T.A.A (Alan) Broadbent	1956–1968

The 'Naval University' was criticized for discouraging practical men because of its academic requirements, and this received some support from an 1877 Committee of Enquiry which stressed that in mathematical studies students without strengths in mathematics should concentrate on topics more suitable to them and of value in undertaking their professional duties. There were other concerns, not surprisingly – preventing cheating, encouraging attendance, and promoting appropriate behaviour. Various recommendations were put in place to address these issues.

Broader curriculum issues also engaged the mathematicians at Greenwich with Burnside's successor Charles Godfrey being extremely influential in the Mathematical Association's teaching committee. He had co-authored an influential geometry textbook for schools which was published in 1903 and sold over 20,000 copies in less than a year.

The twentieth century saw other active mathematical professors and instructors at the Royal Naval College, and it is worth highlighting the College's hosting of the fourth British Mathematical Colloquium in 1952 which, Tony Mann reports, drew the response from the BMC general meeting record that '... the privilege of meeting among such surroundings as these is a notable contribution to the welfare of British mathematics'.

Thomas Archer Hirst was headhunted for the position of Director of Studies which he accepted in January 1873 with a salary of £1,200 and free accommodation. Hirst was born in 1830, and by the time of his appointment to the Royal Naval College he was a senior and respected member of Victorian scientific society. He had been president of the London Mathematical Society, a Vice-President of the Royal Society, Professor of Mathematical Physics, and then Professor of Pure and Applied Mathematics at University College, London. He was an enthusiastic researcher in geometry and had a wide interest in all scientific developments – he was one of the nine members of the influential X-club – as well as a committed and thoughtful educationalist; for example, he was the first President of the Association for the Improvement of Geometrical Teaching (which later became the Mathematical Association). In addition he had travelled very widely in Europe and was well acquainted with the European mathematical community.

Robin Wilson and Helen Gardner draw on many sources for their chapter, particularly Thomas Hirst's diaries which he kept assiduously for 45 years, to discuss his appointment and his impact on the Naval College. The college was to open in October 1873 with 200 students, and before that Hirst appointed the five professors in Mathematics, Applied Mathematics, Physical Science, Chemistry, and Fortification. He visited the dockyard schools to acquaint himself with existing naval education. The syllabuses he designed were in a wide range of subjects for the scientific and technical education of his students. They included calculus, kinematics, dynamics, applied mechanics, physics, chemistry, metallurgy, nautical astronomy, marine engineering, naval architecture, fortification, law, naval history and tactics, modern languages, drawing, and hygiene.

Hirst was an effective and hard-working Director of Studies as well as a capable diplomat in his dealings with the Admiralty since, not surprisingly, areas of difference arose. He also opened the eyes of the College to the wider scientific world with invited lectures from the Astronomer Royal, Airy, and from John Tyndall. The College succeeded with increasing numbers of students and royal recognition when their Royal Highnesses, the Prince and Princess of Wales, presented the prizes on prize day in 1878.

Although his duties were onerous, he did have some time for research, which was so important to him, and Wilson and Gardner discuss this dimension of his life.

During his time at Greenwich Hirst's health was not good, and it was partly for this reason that he resigned his position after 10 years in 1883. The Royal Naval College that he handed on to William Niven was in excellent shape, both academically and administratively, as well as having a national and international reputation.

The Royal Naval College at Greenwich has a rich mathematical history with figures such as T.A.A. Broadbent and L.M. Milne-Thompson, discussed in Chapter 6. But the most influential and productive was the pioneer of group theory, William Burnside (1852–1927), Professor of Mathematics at the Royal Naval College for 34 years from 1885. A number of results in group theory bear his name, and his original statement of what is now known as the Burnside Problem is still not fully solved after over 100 years.

Peter Neumann, in Chapter 8, describes the range of Burnside's research before setting the scene for Burnside's work in group theory, which started relatively late in his life. He does this by initially considering the collection of symmetry operations on an object, for example an isosceles triangle, and then introducing a structure to the collection of symmetry operations, called a group structure. So groups are introduced as a method of measuring symmetry. Restricting to groups with a finite number of elements shows that the study of these finite groups has a strong arithmetical flavour based around the order of the group – which is the number of elements in the group. Using an analogy with the factors of an ordinary integer, the chapter explains how every group has a unique collection of composition factors in the same way that an integer has a unique decomposition as a product of prime factors. A simple group is one which has a particularly straightforward collection of composition factors. With this background established, Neumann is then able to classify Burnside's major interests and results in group theory, relating them to important and exciting work in the subject throughout the twentieth century and up to the present day, and finishing with an interesting discussion of Burnside's interactions from his base in Greenwich with the national and international mathematical community.

0.6 Leslie Comrie and the Nautical Almanac Office

John Pond during his tenure as Astronomer Royal had paid insufficient attention to the production of the *Nautical Almanac*, as a result of which the accuracy of its tables decreased. The national importance given to the Almanac is illustrated by the fact that this decline in accuracy gave rise to questions in Parliament, and in 1818 Thomas Young, Secretary to the Board of Longitude (and translator of the Rosetta Stone), was appointed first Superintendent of the *National Almanac*. In Chapter 9 Mary Croarken begins by describing how Young set about restoring the accuracy of the tables. He concentrated on production of the existing tables and resisted calls from the astronomical community for new tables to assist observational astronomers as well as navigators.

After Young's death the scope of the office was widened following a report by the Astronomical Society, and William Stratford who was appointed in 1832 set up offices in London, abandoning the distributed network of computers. The enterprise returned to Greenwich in 1922 and was based in the King Charles building of the Royal Naval College. The emphasis was still on the improved accuracy of the computing methods and how they could be developed and implemented efficiently. These objectives were to be dramatically achieved under the leadership of L.J. Comrie, and Mary Croarken in this Chapter describes how he mechanized the enterprise of table-making and developed the new discipline of scientific computing.

Comrie joined the Nautical Almanac Office (NAO) in 1925 and was promoted to Superintendent in 1930. From the start he was exploring the use of calculating machines to construct the tables. As a result he needed to develop his numerical algorithms to reflect the capabilities of the machines as well as renew his staff to have the appropriate skills. He was developing ways of making more efficient use of interpolation, and his methods were also pushing the technology to the limits. Comrie seized on each new technological advance, from machines with extra registers to punched card machines. Indeed his work with punched card machines to calculate the motion of the moon was one of the first scientific applications of these machines and was of worldwide influence.

But cutting-edge technology was expensive, and Comrie used his negotiating skills and network of contacts to obtain access to the new (in 1931) National Accounting Machine Class 3000 with six registers – so allowing calculation using sixth differences – a collaboration with the British Mathematical Association Tables Committee. Comrie was assiduous in his dissemination of machine computing techniques within the International Astronomical Union, the British Mathematical Tables Committee, and his own consultancy work. He was a larger than life character and the chapter finishes with a delightful account of his interactions with his Admiralty employers, his dismissal from his position, and his work in retirement.

0.7 The National Maritime Museum and Mathematical Artefacts at Greenwich

In 1823 a National Gallery of Naval Art was founded in the Painted Hall of Greenwich Hospital which also held a separate Naval Museum from 1873. As Richard Dunn describes in his chapter, the origins of the NMM can be traced back to these developments. The NMM opened in 1937 and amalgamated these and other collections, now holding nearly two and a half million items including important and significant examples of scientific, astronomical, and navigational instruments. The term 'scientific instrument' only appeared in the nineteenth century, while the term 'mathematical instrument' can be traced back to the sixteenth century. This chapter takes us on an illuminating journey through the collections, categorizing and relating them to the activities of people working in or near Greenwich illustrating the discussion by reference to particular instruments. One of the major themes developed is the role that Greenwich Observatory played in technical innovation with, for example, the first marine sextant – vital to the solution of the longitude problem by the lunar distance method – on display in the Royal Observatory. Using the instruments in the NMM this chapter illustrates and amplifies many of the new technological developments important to the functioning of the Royal Observatory and introduced in earlier chapters.

The other theme developed is the seemingly more prosaic but no less fascinating one of instruments used on a day-to-day basis for practical purposes. Taking the example of a horary quadrant held in the Museum Richard Dunn describes Greenwich's involvement in planning the change to the Gregorian calendar from the Julian one which happened in England in September 1752, when Wednesday 2 September was followed by Thursday 14 September.

In practical mathematics the NMM's collection of aids to navigation used on board ship is world class. The great storm of late November 1703 claimed as one of its many casualties the 'Stirling Castle' which foundered on Goodwin Sands, off the coast of Kent. Rediscovered in 1979 she yielded a wealth of material used in a ship of the line and in particular a seaman's chest containing practical mathematical instruments – dividers, sand-glasses, the remains of at least one cross-staff, and a Gunter scale. Richard Dunn uses this discovery to illustrate his discussion of mathematical principles in the practical art of navigation and how Greenwich was involved in teaching and preparing boys to become Naval Cadets or Midshipmen.

Another example of a practical mathematical instrument with a rich history, having been used for over a hundred years in surveying, is a station pointer. It was taken on the Polar expedition of 1875–1876. It was then used by the Thames Conservancy, set up in 1857, and subsequently by the Port of London Authority as part of the management of the Thames for the safe movement of vessels until it was retired in about 1980.

As Richard Dunn comments these instruments illustrate the work undertaken in and around Greenwich for over 350 years and are still the subject of study to shed further light on this fascinating collection of the practical tools of mathematics.

0.8 The University of Greenwich

After the Royal Naval College left Greenwich, the Old Royal Naval College became home to the University of Greenwich (from 1999) and, from 2001, to Trinity College of Music (which merged with the Laban Dance Centre in 2005 to become Trinity Laban Conservatoire of Music and Dance). Much of the research of the University's Department of Mathematical Sciences is in the area of numerical mathematics, very much in line with the tradition of mathematics at Greenwich from the work of Flamsteed through that of Maskelyne and Comrie. In Chapter 11 Noel-Ann Bradshaw and Tony Mann trace the evolution of Woolwich Polytechnic into the University of Greenwich, with its commitment to education for practitioners, and its eventual occupation of King William, Queen Anne, and Queen Mary Courts as well as the former Dreadnought Naval Hospital, bringing the story of mathematics at Greenwich up to date and describing some of the mathematics being carried out at Greenwich today.

0.9 Conclusion

Our book finishes with some suggestions of points of interest for a tourist to Greenwich with mathematical interests. The buildings of the former Greenwich Hospital, now the Old Royal Naval College, now house the University of Greenwich and Trinity Laban Conservatoire of Music and Dance. As one leaves the Old Royal Naval College and continues to the River Thames, one comes to the *Cutty Sark*, now resplendent in its new display. As we will see in Chapter 11, mathematics contributed to the survival and preservation for the future of this icon of Greenwich History. The chapters that follow will focus in greater detail on the contributions that Greenwich has made to numerical mathematics and to mathematics in general.

1

The King's Observatory at Greenwich and the First Astronomers Royal: Flamsteed to Bliss

Allan Chapman

CONTENTS

1.1 Introduction

When he came to write the 'Prolegomena', or preface, to his monumental catalogue *Historia Coelestis Britannica* (1725) around 1716–1717, the elderly Revd John Flamsteed bequeathed to us an invaluable history of astronomy.[1] His great preface, occupying 164 folio pages of Latin text, began with the earliest sources then available in classical Greek and Hebrew, the calendars of the Old Testament cultures, and ended with his own researches at the Royal Observatory, Greenwich. What shines through his history, however, is the escalation of astronomical knowledge that had taken place in Europe over the two centuries that preceded his time of writing. Though dealing at some length with the work of those medieval Arab astronomers whose observations had been translated into Latin, he displays an astonishing silence with regard to what modern scholars now know to have been the remarkable astronomical activities of medieval Christian Europe. No doubt, as an Anglican clergyman and a staunch Protestant, he shared the views of his colleague Robert Hooke about the (mythical) 'monkishness' and 'darkness of these times', namely, the Middle Ages.

To Flamsteed, however, 'modern' astronomy began in the late fifteenth century in what was a German-speaking world. For it was figures such as Regiomontanus (Johann Müller) and Bernard Walther who had not only begun to study Ptolemy in the original Greek, but who had made serious observations of the heavens. Walther, for instance, had been the first modern European to do long runs of original observations of the sun, moon, and planets, which he had performed in Nuremberg between 1475 and 1504.[2] And all of these observations had been of geometrical angles subtended between astronomical bodies, from which invaluable positional data could be extracted whereby an astronomer could quantify the precise length of the year and the precession of the equinoxes, construct accurate calendars, and test geometrical models of the heavens.

For while Flamsteed had the highest veneration for brilliant mathematical theorists such as Copernicus, he saw real astronomical progress as data-driven. Between 1510 and 1710, astronomy had become more and more exact in what it could do, and what led this advance were innovations in instrumentation, which enabled astronomers to measure angles with increasing accuracy. For Flamsteed, like most of the astronomers of his day, was a geometer at heart, who saw astronomical advancement as intimately bound up with precision measurement, new data, and prediction. For only in this way could one expect to get to the bottom of the great mysteries of the cosmos, such as the truth or otherwise of the Copernican theory, the physical mechanisms underlying the complex orbits of the moon and planets, and the distances of the stars.

And occupying a pivotal point in what John Flamsteed perceived as his astronomical ancestry was Tycho Brahe. Tycho is pivotal because of his progressive approach to instrumentation: he developed 'families' of instruments, each of which was specifically designed to make one observation to a hitherto unimaginable degree of accuracy. These included quadrants, sextants, armillary spheres, and shadow instruments, whose graduations, sights, and balancing qualities got better and better with each new 'family' member, as Tycho fused classical geometry with Renaissance engineering, so that by 1585, he could measure angles to a single arc minute (or to 1/30 of the diameter of the full moon). And Tycho inspired his Danzig successor, Johannes Hevelius, whose great instruments excelled even those of Tycho in their angle-measuring accuracy. What is more, both Tycho and Hevelius made no secret of their radical new instrument technologies, publishing them, respectively, in sumptuous volumes of fine-art plate engravings with accompanying texts: *Astronomiae Instauratae Mechanica* (1598) and *Machina Coelestis* (1674).[3] And while the foundation of the Royal Observatory, Greenwich, in 1675 was entirely coincidental with the publication of Hevelius's *Machina*, one can fully understand how Flamsteed saw himself in a tradition of great European research observatories, for the precepts and practices of both Tycho and Hevelius were never far from his thinking, and Flamsteed's *Historia Coelestis Britannica* makes it clear that he saw Greenwich as the next development in that tradition.

Where Greenwich did differ from and improve greatly upon Tycho's Uraniborg and Hevelius's Danzig observatories, however, was in the use of telescopes and other devices. For as Flamsteed was all too aware, Galileo's use of the telescope after 1610 had fundamentally altered the astronomical landscape. Yet what was of such importance to Flamsteed was not the discovery of Jupiter's moons or the lunar craters, but the potential application of the telescope to quadrants and sextants, which offered the prospect of being able to measure angles in the sky to single arc-seconds. Tycho, of course, lived just before the telescope was invented, and while Hevelius built long-focus instruments with which to study the surfaces of the moon and planets, he never used them in conjunction with his great brass sextant and quadrants, preferring to make all angular sights by the naked eye alone.[4]

In addition to this mainland European tradition in recent astronomical research, Flamsteed traced the Royal Observatory's history to a particular English source. This lay in Lancashire and Yorkshire, in the 1630s and 1640s, and in particular in Jeremiah Horrocks, William Gascoigne, and William Crabtree, and members of the Townley family, of Townley Hall, Burnley, Lancashire. For Horrocks and Crabtree had observed the first ever recorded transit of Venus in November 1639 and had undertaken fundamental research into testing the Keplerian theory of elliptical orbits; while Gascoigne had succeeded in solving the optical problems involved in measuring small angles through the telescope, inventing the eyepiece micrometer and the telescopic sight. Horrocks, Crabtree, and Gascoigne saw themselves as disciples of Tycho, Galileo, and Kepler, and crucially made significant advances upon their work, which these three north-country astronomers had read in Latin editions in their own homes in north-west England. And while all of these men were dead long before Flamsteed began to take a serious interest in astronomy in the late 1650s, he was able to meet men in Restoration London who had either known them personally, or whose friends and relatives had been acquainted with them.[5]

Central in this respect was Sir Jonas Moore F.R.S., now a pillar of the Royalist establishment and famous as a mathematician and mathematical teacher. But Moore came from pre-Civil-War Lancashire, and even if he had never personally met Horrocks, Crabtree, and Gascoigne (which he could well have done), he had certainly been a client of those scientifically minded Roman Catholic gentlemen, Charles and Christopher Townley, the latter of whom collected up and preserved their surviving scientific papers, for all three – plus Charles Townley – were dead by 1644. (Both Gascoigne and Charles Townley were killed on the Royalist side in the battle of Marston Moor in July 1644, though Christopher survived.) In 1672, Flamsteed made a personal visit to Townley Hall, where he not only looked through and transcribed sections of the three north-country astronomers' papers, but even saw the 'iron carkasse' of Gascoigne's sextant.[6] And it had been the current head of the family, Richard Townley (Christopher's nephew), who in 1667 had communicated a design for a telescopic eyepiece micrometer, based on William Gascoigne's original, to the *Philosophical Transactions* of the Royal Society.[7] So great was Flamsteed's regard for Crabtree, Gascoigne, and Horrocks, and so formative did he see their role in the creation of British astronomy that he began Volume I of *Historia Coelestis Britannica* by printing six folio pages of Gascoigne's and Crabtree's observations, made between 1638 and 1643.

So here, we see the intellectual lineage and inspiration of the first Astronomer Royal. For Walther, Tycho, Hevelius, and others brought home to him the power not only of ideas in the transformation of astronomy but also of instruments, techniques, sound engineering structures, and refined practice. Then to these must be added Horrocks, Crabtree, and Gascoigne and their circle, for patriotism also played its part in his scientific vision. Not a blind patriotism, indeed, but a clear perception of what his fellow countrymen had done, and of his duty to build upon the foundations they had laid. For let us not forget that Flamsteed's professional career as an astronomer coincided almost exactly with the reign of King Louis XIV of France, during which time England and France were locked together in a succession of wars in which King Louis attempted a French political and military domination of Europe. Indeed, English genius was to be celebrated, especially as continental giants such as Hevelius were already publishing Horrocks's manuscripts at their own expense. And Christiaan Huygens carried his own praise of their achievement into the Dutch Republic.[8]

But central to everything, as far as Flamsteed was concerned, was accuracy and its improvement. For John Flamsteed was a deeply practical astronomer, whose real concern was the investigation and elucidation of nature, prior to its practical application to navigation and such. Abstraction and love of theory for its own sake were never part of his intellectual architecture. The great questions in current astronomy would be solved by making better tools – more accurate instruments – with which to tackle them, and this deeply functional approach to astronomy which Flamsteed epitomized would become the hallmark of Greenwich science down to the twentieth century.

One might also suggest that this hands-on pragmatism in astronomy was a reflection of John Flamsteed's theology as an Anglican parson, for nature and the heavens were the touchstones of God in the created world in the same way that the Bible was the touchstone in the world of the spirit. For John Flamsteed was a deeply devout Protestant astronomer who believed that truth was there to be known if only one sought it in the right way.

1.2 The Founding of the Royal Observatory

The circumstances which led to the founding of the Royal Observatory in the summer of 1675 are well known. A French mathematical adventurer, Le Sieur de St. Pierre, had approached King Charles II under the auspices of his current French mistress, Louise de Kérouaille, with a scheme for finding the longitude at sea by means of measured sightings of the moon and certain bright stars. Indeed, the idea of 'lunars' already went back over a century by 1675, and St. Pierre's method was only a variant of other unsuccessful ones. St. Pierre's proposal, however, was handed over to the newly established (1660) Royal Society of London for an expert opinion. Flamsteed himself had not yet been elected to their Fellowship and was still living with his father in Derby, but it happened that he was in London on business when St. Pierre's proposal was being talked about. Because Flamsteed already had friends in the Royal Society and enjoyed a reputation as a learned and skilled practical astronomer, he was invited to give his own report on St. Pierre's method. He pointed out its derivative nature and told the King that St. Pierre's astronomical knowledge was not good; otherwise, he would have been aware that the crucial tabular data that his theory needed to make it work did not exist. Quite simply, none of the available published astronomical tables could be used to predict the future position of the moon amongst the fixed stars – fundamental to St. Pierre's longitude method – to anything like the necessary level of accuracy to establish the position of a ship on the high seas. His Majesty expressed concern for the safety of his brave sailors and asked what would be necessary to develop a reliable method of finding the longitude. Flamsteed said that nothing short of a fundamental re-observation of the moon's orbit amongst the stars would do: a fundamental re-observation, moreover, conducted with the latest and most accurate instruments.[9]

By 1675, John Flamsteed was very much aware that a radical improvement in angular measuring accuracy had taken place even within his own 29-year lifetime. For had not Gascoigne's successful application of telescopic sights to large-scale sextants and quadrants 7 or 8 years before he was born enabled stellar angular separations to be measured 40 times more accurately than with the naked eye? And did not Gascoigne's micrometer, placed inside the eyepiece of a refracting telescope, make it possible to measure the angular diameter of the sun or moon on any given occasion to 2 or 3 arc-seconds? For as Gascoigne himself had realized, if one could measure daily changes in the diameter of these bodies, then one could construct the elements of the elliptical orbits of the earth around the sun and of the moon around the earth, to a new and unprecedented degree of accuracy. Such data, indeed, would be essential to finding the longitude, for while the lunar longitude method was itself hands-on and practical from the sea-going navigator's point of view, it could only work if it stood on a firm foundation of pure science research, practical, observational, and mathematical. This understanding was central to Flamsteed's thinking, and without it the Royal Observatory would never have come into being.[10]

And just as the telescopic sight and micrometer had revolutionized the quality of direct angular data, so the invention of the pendulum clock in 1658 had revolutionized time-keeping, especially as Christiaan Huygens's original escapements were improved upon in London. Indeed, the original Dutch clock escapements could only work with short pendulums swinging through very wide angles, which in turn made them subject to air resistance, secondary vibration, and other problems. But by 1670, Robert Hooke and William Clement in London had devised an improved escapement which could work with a swing of no more than five or so degrees of arc, and could therefore control much longer pendulums. The physics of the pendulum had become a topic of intense study in the late 1660s and early 1670s, when it was found that longer pendulums were slower and more stable in their vibrations than were short ones. A pendulum of 39.25 in long, for instance, was discovered to swing exactly in one single second of time in London, while one of 13 ft took two seconds of time to accomplish one swing. Such pendulums brought in a revolution in time-keeping accuracy that was not to be bettered in pure physics terms until the electrically excited quartz crystal with its even more exact vibrations made its appearance in the mid-twentieth century.[11]

Flamsteed immediately grasped the potential of such clocks. They could, for instance, be employed to confirm the homogeneous rotation of the earth. First, one could use successive daily transits of a bright star, such as Sirius, observed with a fixed telescope to establish the exact duration and homogeneity of the sidereal day. Then, second, one could use this fixed standard to quantify the seasonal speeding-up and slowing-down of the sun in the meridian (caused by the earth's elliptical orbit and changing velocity in accordance with Kepler's laws) to quantify the solar year and hence fix a mean velocity for the Sun as it appears in the sky, to work out mean solar time at Greenwich. Without a clear and demonstrated knowledge of the homogenous rotation of the earth, indeed, it was hopeless trying to use lunar angles as a way of finding the longitude, for both astronomers and navigators on the high seas had to rest secure that their respective tables and observations shared a common, uniform, time-scale.

Indeed, Flamsteed so impressed the King and the Royal Society that His Majesty decided upon the establishment of a Royal Observatory in Greenwich Park, high up on the hill, overlooking the Queen's House, the Kent Road, the docks, and that great sweep of the Thames that bore much of the world's trade and shipping up into London. Unlike King Louis's Paris Observatory, Greenwich was not intended to be a show-piece of state. Instead, the King's architect, Sir Christopher Wren, was given a limited budget of £500. The foundation stone was laid on 10 August 1675, and the roof was on by Christmas. Indeed, it was economy architecture of the very highest order. From start to finish, it cost only £520–9s-1d, just slightly over budget, and was constructed from second-hand building materials, reclaimed from the recently demolished fort at Tilbury. These in turn were paid for by the proceeds of the sale of 'decayed' gunpow-der in the Tower of London arsenal, the powder itself being recycled, with its sulphur and saltpetre being extracted by chemical manufacturers.[12]

Yet from these skimped resources, Wren was to produce one of the most beauti-ful small buildings in the whole of seventeenth-century England. The Observatory rose up on the site of old Greenwich Castle, on a ridge of raised ground, command-ing spectacular views of the wide sweep of the Thames as it twisted on its way into London, and looking northwards across the Essex marshes. And being 6 miles east of London in that un-light-polluted age, it would have had skies as dark as any in southern England. John Flamsteed was designated, in his Royal Warrant of 4 March 1675, as 'our Astronomical Observator', at a salary of £100 per annum. This sum, on paper at least, was not bad for the time. The snag lay below the surface: Flamsteed not infrequently had to go to considerable lengths to extract his quarterly payments from His Majesty's rather chaotic administration, added to which he was expected to find his own instruments and equipment, not to mention any assistance that he might need beyond the services of his officially provided assistant, who received only £25 a year. By the winter of 1675–1676, however, John Flamsteed had an empty Observatory building, a modest yet adequate on-site residence, an official salary, and a title that would soon be known internationally, and for the following centuries, as 'Astronomer Royal'. Although not a university graduate, Flamsteed was given the degree of M.A. by Cambridge University and was ordained into the Diaconate of the Church of England by the Bishop of Ely at Easter 1675, and priested in 1684. As the Revd John Flamsteed, M.A., Cantab., it was hoped that he would also receive an ecclesiastical benefice to augment his stipend as Astronomer Royal. This preferment did not materi-alize until 1684, when Flamsteed became Rector of Burstow, Sussex. Yet as he made clear to the visiting German scholar, Zacharias Conrad von Uffenbach, in 1710, his achievement at Greenwich would have been impossible had he not been 'the son of a rich merchant', for following the sudden death of his father Stephen Flamsteed, a Derby brewer, in 1688, John, the Astronomer Royal, acquired the personal resources to commission major instruments of research.[13] Indeed, Flamsteed later claimed to have spent over £2,000 of his own money in equipping the Royal Observatory with a progressive series of research tools so that he could carry on the traditions of Tycho and Hevelius and make advances beyond them.[14] So, one might ask, with this colos-sal level of expenditure – hundreds of thousands of pounds in modern money – was John Flamsteed, a paid professional astronomer, or a member of that class of English 'Grand Amateurs' who undertook fundamental astronomical research at their own expense? Indeed, one can fully understand why Flamsteed was so furious when, by 1710, Sir Isaac Newton was making patronising remarks about his work at Greenwich. For in reality, John Flamsteed was a scientific civil servant who was obliged to pay his own way.

1.3 Flamsteed's Instruments

Central to John Flamsteed's entire achievement at Greenwich was a new approach to instrumentation. Mention has already been made of the technological innovations which had taken place in instrumentation since 1640, with the telescopic sight, micrometer, and pendulum clock, and Flamsteed saw access to instruments possessing such refined angle-measuring capacities as being fundamental to his work in the Royal Observatory. Having no grant with which to equip the Observatory, however, he was forced back onto his own resources. As he seems to have owned several instruments, including telescopes, which he had at his father's house at Derby, he brought this private collection down to the Observatory. The Royal Society also loaned him some rather experimental pieces from its own collection, including a couple of novel screw-edged quadrants devised by Robert Hooke. Indeed, we have a list of these pieces, for in 1679, Hooke re-took possession of them. But absolutely central to what he would do was the friendship and support of Sir Jonas Moore, the Lancashire astronomer who formed the living link between the three north-country astronomers and their friends of 1640 and the new Royal Society. As Master of the King's Ordnance, with private money to spend and access to the skills of the armourers and blacksmiths in the Tower of London, Sir Jonas, more than any other single individual, was instrumental in getting the Royal Observatory up and running. He was indeed a munificent patron to Flamsteed. Out of his own pocket, Sir Jonas provided Flamsteed with clocks of the latest design and accuracy, made by Thomas Tompion. He also commissioned, on Flamsteed's design, an iron-framed sextant of 7 ft radius, equipped with telescopic sights. Robert Hooke's micrometer adaptation of Gascoigne's design for reading tiny angular fractions was incorporated, and all set upon an iron equatorial axis. Now Flamsteed and an assistant could measure the angles between any given pair of astronomical bodies and cross-check their accuracy through a series of interlocking celestial triangles. This instrument clearly drew its inspiration from the large sextants of Tycho and Hevelius, but was vastly improved, and incorporated new and refined technologies.[15]

But as the sextants measured the 'longitude', or east-west, angles between objects, Flamsteed also needed an exact meridian quadrant to act as a fixed reference point in the heavens. This quadrant could not only measure the altitudes – or declinations – of the sun, moon, and stars as they came due south, but Flamsteed also hoped to use it as a meridian transit point against which to make timings from his clocks. Robert Hooke, who in 1675 was at the peak of his creative brilliance as a designer of new research instruments, produced a plan for an enormous quadrant of 10 ft radius – 4 ft bigger than Tycho's prototype of 90 years before – that would be iron-framed, secured to a meridian wall, and equipped with a telescopic sight and a novel micrometer of Hooke's own devising. It was by far the most advanced quadrant ever designed, as was the sextant for that class of instrument. Sir Jonas seems to have paid for both out of his own pocket in an act of wonderful public-spiritedness, using the Tower of London craftsmen to execute the sophisticated ironwork and employing the clockmaker Thomas Tompion for all the precision parts.

The sextant, after overcoming teething troubles, was an enormous success and would give decades of sterling service. Hooke's 10-foot quadrant, alas, proved an expensive failure. Its frame was too weak for its great weight. Hooke's novel micrometer, geometrically elegant as it seemed on paper, turned out to be a flop in practice, and the overall mechanism required large radial arms of iron to scissor past each other in a way so dangerous that it almost, as Flamsteed recorded, 'deprived Cuthbert of his fingers'.[16] (This was Cuthbert Denton, Flamsteed's assistant.)

Both instruments, along with the Tompion clocks, seem to have come into use in 1676, and though still lacking an effective and reliable quadrant, John Flamsteed applied himself with great energy to triangulating the heavens and recording the moon's complex orbit amongst the fixed stars – from which, it was hoped, tables of such accuracy would ensue as would enable the longitude to be 'discovered'.

Jonas Moore also seems to have paid for a pair of clocks of novel design for Flamsteed, with pendulums 13 ft long, moving through tiny angles of oscillation and communicating with their gear trains through ingenious 'dead beat' or recoil-free escapements designed by Richard Townley.[17]

Flamsteed developed a research procedure with these instruments which, though deriving in its basics from Tycho, was now capable of recording celestial angles to a few arc-seconds for any given observation.

To map a constellation, let us say Orion, this is what he would do. The quadrant – Hooke's or a smaller instrument – was used to take the vertical or declination angle of each star, as the constellation moved across the meridian, in a westbound direction over about 45 minutes of time. First would come Rigel, then Bellatrix, then the stars of Orion's belt, then Saiph, and lastly Betelgeuse, along with a cluster of dimmer stars. On a clear winter's night, Flamsteed might get all the stars in the constellation during a single meridian passage.

Next, he would use the great 7 ft radius sextant, on its novel equatorial mount. Being mounted on a flexible iron pivot head, the flat plane of the sextant could be adjusted to any desired place in the sky, and with Flamsteed directing the first of the instrument's 7 ft telescopic sights to, let us say, Rigel, an assistant could rotate the second pivoted sight and set its crosshairs upon Betelgeuse. When both observers were confident of the adjustment of their respective telescopic sights, the engraved scales of the sextant, used in conjunction with its screw micrometers, could be read off to yield the exact angle between the stars. In this way Flamsteed could not only map all the star positions in Orion, but could do the same for every constellation in the sky, in a series of great interlocking celestial triangles. And when the height, or dec- lination, angles from the quadrant were added, a map of the sky, drawn to a hitherto unimagined accuracy, began to emerge. And once the zodiac pathway around the heavens – through which the sun, moon, and all the planets move – had been mapped to the desired standard, Flamsteed could record the nightly positions of the moon and planets, and of the sun by day. For central to Flamsteed's remit as Astronomer Royal, and the reason why King Charles had founded the Royal Observatory in the first place, had been the need to produce superior tables of the moon's motion, from which a navigator might find his position at sea.

Flamsteed never lost sight of this remit, and while over the 45 years that he was in office as Astronomer Royal, he undertook researches into timing the earth's rotation to see if it was homogeneous, attempted to measure stellar parallaxes, and engaged in other projects, all were seen as ancillary to the perfecting of navigation.

By the late 1680s, the gift of an ecclesiastical benefice, Burstow, Sussex, and the inheri- tance of his deceased father's Derbyshire money had made John Flamsteed a comfortably off gentleman. Being able to measure east-west angles (right ascension), more reliably with his stalwart sextant than he could in the vertical (declination) with his unreliable quadrants, Flamsteed decided to spend £120 of his own money – or some 14 months' worth of his official salary – on commissioning a new meridian instrument that would match the sextant in dimensions, graduations, rigidity, and reliability. It was built – or probably designed – by his one-time friend and assistant, the Yorkshire astronomical country gentleman Abraham Sharp. And as soon as it came into commission, in 1689, Flamsteed realized that, at last, he had the quadrant of his dreams. For its rigid 7 ft radius iron frame, delicately engraved brass degree-scales, micrometers, and telescopic sights made it the most accurate angle-measuring instrument in the world. Its scales read directly down to 10 arc-seconds, and hopefully less when using micrometers.[18]

And strictly speaking, Flamsteed's new instrument was not a quadrant, but what he called a 'Mural Arch', encompassing a full 140 degrees of the meridian. The purpose of encompassing this vast sweep of sky was to enable Flamsteed to swing his telescopic sight around from the southern to the northern meridian, where he could observe Polaris, the Pole Star, and make a critical determination of the latitude of the Greenwich Observatory.

Once this 'Mural Arch' had been successfully brought into operation, Flamsteed found that it provided a far more convenient, and rapid, method by which he could measure the east-west or Right Ascension angles of stars than did the big two-man sextant. This was done by using the Arch in conjunction with a new clock commissioned by Flamsteed, also from Thomas Tompion, in 1691, which was designed not so much to tell the time in the normal way as to mark each degree of Right Ascension, as it culminated in the meridian. So all that Flamsteed needed to do to measure the angle between any two objects in the sky was to make an exact timing, to at least a second, of how long it took each object to cross that meridian. As the 1691 Tompion clock was regulated to keep time by the stars, and not the sun – to mark the sidereal day of 23 hours 56 minutes and 4 seconds – Flamsteed could now use his clock to reliably measure angles. Durations of time now yielded their exact equivalents in degrees, minutes, and seconds of arc.[19]

Amongst other things, this new way of working by Mural Arch and clock after 1691 had a salutary effect on the Astronomer Royal's often uncertain health. No longer did he need to work in an exposed, roofless building, in a stiff icy wind, to measure Right Ascensions with the sextants, for the Mural Arch was entirely enclosed within a building, and access to the meridian sky was obtained by the simple expedient of opening a 6-inch-wide shutter in the roof. The 'tortures' of the gout, the stone, headaches, neuralgia, and chills which the cold night air brought on for Flamsteed were now largely avoided, and while he did continue to make use of the sextant, his observing logs make it abundantly clear that after 1691, the Mural Arch had become his major observing instrument for both right ascensions and declinations.

But why did Flamsteed keep needing to observe, re-observe, and observe yet again the stars and planets, over nearly half a century of ceaseless toil? The reason lies in his relentless quest for perfection. The values for given observations and positions had to be constantly refined and reduced until the position of every object in the sky was reliably logged to a small fraction of an arc-minute. And just as his hero Tycho had tried to attain perfection not only through ceaseless re-observing and refining, but also by endlessly cross-checking an observation made with different instruments, such as his quadrant, sextants, and armillary sphere, so Flamsteed aspired to do the same. The sextant, Mural Arch, and clock would be used in turn to measure the same angles in a continuous process of refinement. By this process, Flamsteed would not merely map the heavens and in 45 years observe the moon through more than two of its great complex 18-year cycles; he would use this growing body of data to extract a series of astronomical 'constants', such as the First Point of Aries, the annual rate of the precession of the equinoxes, the exact duration of all the mean solar and sidereal years, and several other things besides. These would provide the essential tabular and cartographic data on which a system of practical astronomy could be built, which would be an essential predicate to understanding the lunar orbit so well as to be able to use it to find a ship's longitude at sea. So Le Sieur de St. Pierre's opportune ignorance about what the longitude really entailed had not only created a major national and international institution – the Royal Observatory – but had helped to set astronomy on an entirely new footing of accuracy, backed by the new telescopic sights, micrometers, and pendulum clocks.

Looking through Flamsteed's observing books, now preserved as part of the Royal Greenwich Observatory manuscripts in Cambridge University Library, one becomes aware of the enormous labour of love which the first Astronomer Royal bestowed upon his institution. A labour of love in that he was inadequately paid, and by necessity self-funded as far as instrumentation, administration, and anything beyond the most rudimentary assistance was concerned. It was all the more gruelling, therefore, when the well-meaning Prince George of Denmark, Queen Anne's husband, tried to set in motion a project for the publication of the observations of John Flamsteed in 1704. For this afforded an opportunity for the newly knighted Sir Isaac Newton, President of the Royal Society, to interfere.

It is quite clear from their correspondence and other writings that Newton and Flamsteed were not friends. Newton seems to have been unconcerned about Flamsteed's *de facto* self-funded status and complained that he had been dilatory in presenting his observations to the public. And as Flamsteed's manuscript observations were the finest ever made, Newton also wished to get his hands upon them as a way of checking his lunar theory, as understood within his new theory of universal gravitation. As part of a 1705 legal agreement regarding the publication at royal expense, Flamsteed was obliged to hand over his observing books to Newton as a committee, under Newton's direction, took over the work. And once this had been accomplished, the presses seemed to stop or at best only to creak along. Indeed, a not entirely generous person might suggest that as Newton, at last, had his hands on the data that he wanted, he rather lost interest in the publication project. Finally, in 1712, under the editorship of Edmond Halley and containing a 'Preface' that was highly critical of Flamsteed, an uncorrected version of his Greenwich observations was published in two sumptuous folio volumes entitled *Historia Coelestis*. The meticulous Flamsteed was, predictably, apoplectically furious.[20]

Following the political confusion after the death of Queen Anne – who died without heirs in 1714 – Flamsteed was able to lobby friendly politicians to have returned to him the unsold copies of his 'pirated' and unofficial observations. The letter which he wrote to the Vice-Chancellor of Oxford University requesting – unsuccessfully – the handing over of the 'pirated' volumes from the Bodleian Library survives and captures his anger over what had happened.

Then, in 1715, in his 69th year and far from well, Flamsteed set about publishing his monumental life's work, at his own expense. It is true that some sheets of observations from the pirated edition were good enough to use, but he made a bonfire of the surviving pages: a sacrifice to heavenly truth, as he styled their burning. He now employed his old assistants, Joseph Crossthwaite, James Hodgson, Abraham Sharp, and others, to re-work his raw data: men he knew well and could trust to be accurate and honest. He also appointed printers and then had the sheets sent to proof-readers whom he knew and who lived in different parts of England, and who clearly could not collude to do a scamped job. His old Derbyshire connections are evident in his employment of a Mr. William Bosseley, an apothecary of Bakewell in that county, and of the neighbouring Mr. Luke Leigh, as his, presumably paid, checkers.[21]

He then set about writing the massive, book-length Prolegomena mentioned above. Here, in the English manuscript drafts, which survive in Cambridge University and the Royal Society Libraries, Flamsteed not only told the history of astronomy, his sense of the Tychonic tradition, and the Royal Observatory's place in that tradition as it was then understood, but also related his treatment by Newton, and the behaviour of Newton's 'creatures' and acolytes to 'cry up' to him, and how the now great Sir Isaac would do his best to marginalize anyone who stood up to him. It was a pretty powerful story, but it did not see the light of day until 1982, for the Revd John Flamsteed died at 9.30 p.m. on the night of 31 December 1719, leaving his masterpiece unfinished.

It was Margaret Flamsteed, the Astronomer Royal's formidable widow, with her business ability and determination to see her husband's work made public in its proper form, who took over the project, and she now started to pay Hodgson, Crossthwaite, and the others out of her own pocket. She appointed printers and supervised the whole vast enterprise to final publication in three magnificent folio volumes, the *Historia Coelestis Britannica* (*The British Account of the Heavens*) in 1725 (Figure 1.1). Of course, following John's death, Margaret Flamsteed was not only obliged to vacate the

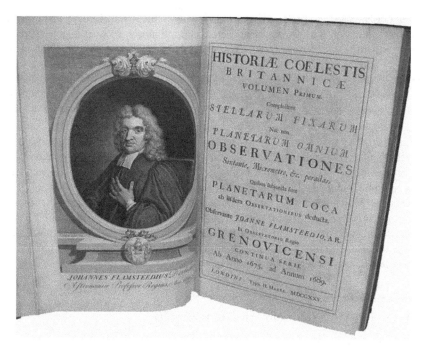

FIGURE 1.1   Title page and engraved illustration of John Flamsteed: *Historia Coelestis Britannica* (volume 1). London: 1725. (Courtesy of Special Collections, Glasgow University Library.)

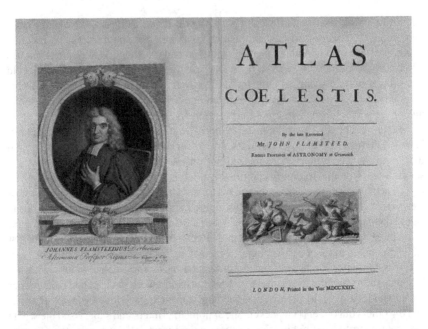

FIGURE 1.2   Title page of Flamsteed's *Atlas Coelestis*, 1729. (From Internet Archive https://archive.org/details/atlascoelestis00flam/page/n5.)

Royal Observatory and find, and pay for, her own premises, but she even had to fight off a lawsuit in which the government tried to claim Flamsteed's instruments given to him personally by Sir Jonas Moore, to establish them as his own and not government property. She won, though unfortunately they were then sold and subsequently lost. And in volume 3 was printed Flamsteed's Monumental 'Preface', or 'Prolegomena' as it appeared when translated into Latin, and aimed, rightly, at an international scholarly market. Sadly, however, Flamsteed's posthumous editors and translators, knowing, no doubt, that they would have to find work elsewhere when the great *Historia* was delivered to the world, carefully removed the blunt home-truths about Newton which Flamsteed had put in his English draft. Had Flamsteed lived long enough to see his *magnum opus* delivered to the world, in the form that he intended, one wonders how the subsequent beatification of the memory of Newton might have proceeded![22]

Then, all was brought to a fitting conclusion in 1729, when Margaret Flamsteed and her team of calculators, artists, and engravers brought out the spectacular *Atlas Coelestis* on thick sheets of 'Elephant Folio' paper, the biggest sheets that eighteenth-century presses could handle. For here, all the star positions were turned into great constellation maps. Science and art were combined! (Figure 1.2.)

1.4 Paying for Astronomy in Great Britain

We should not forget that very little British astronomical research of the most innovative kind was done at public or government expense before the mid-twentieth century. It was the private scientist who made the greatest contributions and dug deep into his own pocket to provide the instruments and other resources. And even the first six Astronomers Royal were true to this pattern. For not only did Flamsteed dig deep into his own pocket, both to observe for and to publish the *Historia*, but all of his successors until Sir George Airy, who was appointed Astronomer Royal in 1835, were expected to be gentlemen of means. For while after 1720, the government formally acknowledged its duty to supply the Astronomer Royal's instruments and a few other necessities, the salary was to remain modest, given his status, until the Victorian age. For an Astronomer Royal was expected to be a 'proper' gentleman, whose time was not wholly paid for by the government and whose 'official' papers remained his own property, to be removed by his executors from the Observatory when he died. And even if he did not enjoy inherited private means as such, then he was expected to be a beneficed clergyman, a pensioned Post Captain in the Royal Navy (like Halley), a semi-resident professorial Oxford don, 'timesharing' his Greenwich work (like Halley, Bradley, and Bliss) with a senior academic teaching post, or married to a wealthy heiress (like Bradley and subsequently Maskelyne). And some of these later holders of Flamsteed's Office enjoyed multiple revenues in addition to their Greenwich job. James Bradley, for instance, was a clergyman, a successful Oxford professor (offering lucrative extra courses) as well as being the husband of Mary Peach, a Gloucestershire woollen milling heiress. With advantages like these, the government could keep the official salary modest, and in 1810, a requested pay rise for the 78-year-old Nevil Maskelyne was turned down flat on the grounds that, while he had not indeed had a pay rise for 45 years, Dr Maskelyne was a rich gentleman who could not expect to make any further demands on the public purse![23] It was not until the hard-up scholarship boy Airy told the appointing Government and Admiralty Commission in their negotiations up to 1835 that he needed to be paid a viable professional salary, for he was not a clergyman or a naval officer and had no private means at all, that a realistic stipend came to be paid to the Astronomer Royal and his staff. (Airy had negotiated similarly with Cambridge Senate upon being appointed Professorial Director of the University Observatory in 1828.)[24]

Flamsteed's great *Historia* set a new standard for astronomical research that would provide a benchmark for almost a century to come, and while new levels of astronomical accuracy were already outstripping his measured values even by 1725, Flamsteed's vast assemblage of data would inspire astronomers across Europe for several generations to come and supply a ground plan for Sir William Herschel's cosmological researches after 1780.

Yet as it turned out, even Flamsteed's meticulously conducted angular measurements were not sufficiently accurate to pin down the lunar orbit to the degree of precision that would be necessary to find the longitude.

1.5 Edmond Halley

A career like that of Edmond Halley (Figure 1.3) would not be possible in the present day and age. For one thing, at 64 years old at the time of his appointment as Astronomer Royal, he would be deemed far too old for office. And as a man with such an exotic career background – independent gentleman, traveller, Clerk to the Royal Society, inventor, entrepreneur, Royal Navy Captain, deep-sea diver, Oxford professor, diplomat, inspired gadfly of the physical sciences with several score of

FIGURE 1.3   Edmond Halley. (Reproduced from E. Walter Maunder, *The Royal Observatory Greenwich: A Glance at Its History and Work* (London: The Religious Tract Society, 1900).)

papers to his credit in the Royal Society's *Philosophical Transactions* – he would perhaps now be seen as too 'unfocused' to please a modern high-powered government committee. For without a doubt, there really was something larger than life about this London merchant's son, who had broken off his undergraduate studies at Oxford to sail to St. Helena to map the stars of the southern hemisphere, and received his M.A. degree without examination, and by Royal Command, at the age of 22. And what a contrast to Flamsteed as a personality type! For Halley was as hard as nails, yet jolly and good-natured, and eminently untroubled by illness, *angst*, or the state of his soul. He was also a close friend of Newton. Indeed, it was Halley, more than anyone else, who had possessed the sheer cheek and thick skin to gently wheedle the touchy and often paranoid Newton into writing *Principia Mathematica* in the first place, and even ventured his own time and money in seeing Newton's great work through the press.[25]

It was Flamsteed who had given Halley his first lessons in practical observational astronomy with telescopic sights, micrometers, and pendulum clocks, prior to his sailing to St. Helena in 1676, though his close friendship with Newton, and the not entirely honourable part which he played in editing the 'pirated' version of Flamsteed's observations in 1712, made Halley out to be Newton's arch 'creature' by the 1710s. And looking at things from Flamsteed's perspective, one can fully sympathize.

So when Halley was appointed Astronomer Royal in 1720 and moved into a set of buildings stripped bare by his predecessor's executors, one has little difficulty in understanding Margaret Flamsteed's fury, for Halley's appointment must have seemed like officialdom's parting insult to the memory and achievement of her late husband. Nor can we blame her and others of Flamsteed's friends for seeing the hand of Newton in the appointment. What must have seemed like supreme over-confidence, however, was Halley's declared intention to observe the moon throughout its entire 18-year cycle: an undertaking which would keep him hard at it into his early 80s. And as the Longitude Act of 1714 had laid down a £20,000 prize for anyone who could fix the longitude of a vessel at sea to within quite close geographical parameters, Halley intended to use the moon to help him win the prize. Indeed, the gentlemanly status of the Astronomer Royal, as opposed to what one might think of as his Civil Service employee status, is made manifest in the fact that he clearly regarded himself as a free competitor for the prize. After all, he still only received Flamsteed's £100 a year salary, topped up with a quite independent Royal Navy pension or Captain's half-pay, so surely he was a private gentleman holding a public office and living in an official residence! Employment was viewed differently in the eighteenth century, if one were fortunate enough to be a gentleman.

Where Halley differed significantly from Flamsteed was in the way in which he acquired new instruments for the Royal Observatory. Keeping his own money firmly in his pocket, he extracted a hefty grant of £500 out of the government of King George I. Of course, what he wanted was a kit that was essentially similar to Flamsteed's, with a large quadrant for observing transits and declinations but a new type of instrument – a transit – with which to measure Right Ascensions from the meridian, by means of a precision 'astronomical regulator' clock. Halley's shrewdness created a vital precedent, for from 1720 onwards it was recognized as the government's job to provide and maintain the Royal Observatory's instruments, and not the financial responsibility of the Astronomer Royal. And of vital significance for historians of science, this meant that instruments remained in the Observatory's possession, even after technological advancement had rendered particular pieces redundant. This is one reason why today the old instruments, preserved now on site in the National Maritime Museum, are such an invaluable historical resource.

1.6 Halley's Instruments

Flamsteed's 44 years at Greenwich had seen a revolution in precision technology, predominantly in London, as clockmakers, scale-dividers, precision screw-cutters, glass-grinders, and others had transformed the technological landscape. And of course, Flamsteed's Greenwich had played a flagship role in that wider process. By 1720, however, big sextants, such as those used by Tycho, Hevelius, and Flamsteed, had become museum pieces, along with armillary spheres. By 1720, indeed, Halley realized that one could do pretty well everything in practical celestial geometry with three instruments: the transit, quadrant, and clock.

A transit instrument was a simple refracting telescope, mounted on a pair of trunnion supports, somewhat resembling a cannon, and rotating in the meridian on a pair of self-centring 'V' bearings. Such an instrument had no graduated scales or any other specifically mathematical parts. Its sole purpose was to provide a flawlessly trustworthy meridian mark, through 180°, and passing through the north celestial pole. In conjunction with an excellent pendulum clock, it could be used to measure the east-west Right Ascensions of stars by timing the intervals between which any two objects crossed the meridian, and converting the ensuing clock time into an angle as a fraction of 360°.[26]

To measure his vertical, or Declination, angles, Halley ordered a great iron-framed and brass-limbed quadrant of 8 ft radius.[27] It was a triumph of engineering in its own right, while its limb was engraved with two distinct sets of scales: one based on the conventional 90 degrees, and the other parallel to it and an inch away, divided into 96 equal parts. Drawn in accordance with two different sets of geometrical principles, the two scales were designed to act as a cross-check upon each other: when both were in agreement, Halley could be confident that he had an exact, error-free angle. When they differed slightly, a pre-calculated and pre-tabulated correction would be applied.

The transit and quadrant, along with their companion regulator clock, were all the work of one man: George Graham, who was Tompion's old pupil and who inherited the great man's mantle as horologist and precision engineer. And as if to confirm Graham's status as a scientist in his own right, in addition to being a craftsman, he was also a Fellow of the Royal Society and a good friend of Halley. He was the pattern, in fact, for several craftsmen-scientists in the eighteenth and nineteenth centuries, who were sometimes even Fellows of the Royal Society, and whose careers demonstrate what an open world British science was becoming. A world indeed in which intelligent, bench-trained master-craftsmen, who had mastered practical mathematics and shown a genius for experimental investigations financed by their own business enterprises, could join the nation's scientific élite as equals: a thing which was to amaze visiting French *savants*.

Halley entered into office as Astronomer Royal as an extremely vigorous 64-year-old who had a formidable track record behind him, not just as an observational astronomer with a superb mastery of practical instrumentation but also as an analytical mathematician, cometographer, organizer, far-sighted visionary, and even a kind of popularizer. Most of all, he had a reputation as a man of action in every sense: a man who got things done. As a visionary, as it were, he had been aware, since viewing the 1677 transit of Mercury from St. Helena, that the forthcoming transit of Venus in 1761 could provide a crucial opportunity for measuring the solar parallax and re-defining the distance of the sun. It was visionary in so far that no one who heard his clarion call in 1690, or even in 1716, was ever likely to personally see the Venus transit with their own eyes. His Latin papers on how the transit might be observed were intended as a wake-up call to future generations. Likewise, his publications in advance of the total eclipse of 1715, which would pass across England, were intended to alert, and recruit, everyone who owned a decent clock or a telescope. Dozens of people sent their results to him – country gentlemen, clergy, the Bishop of Exeter, and even a tax collector – all of whom he named and acknowledged in print. Perhaps the first coordinated 'mass observation' of an astronomical event on record. Halley himself observed the eclipse with a friend, Lord Macclesfield, who was a senior judge in addition to having scientific interests.[28]

To see Halley, therefore, as nothing more than Newton's place-man, getting the post of Astronomer Royal undeservedly, is to fundamentally misunderstand his colossal standing in the world of science in 1720. For he was seen as ranking second only to Newton as a 'genius'. And added to the above, he was Oxford University's Savilian Professor of Geometry (he had been an undergraduate at The Queen's College, Oxford), a post which he was to hold in tandem with that of Astronomer Royal right down to his death in 1742.

Incredible as it may appear, Halley was actually 69 years old by the time that George Graham had completed the last of his state-of-the-art set of instruments, the 8-foot quadrant, in 1725. Yet in spite of his efforts, even Halley could not obtain values of sufficient accuracy to make it possible to find the longitude by lunars. Keen to win the £20,000 Longitude Prize as he was, however, Edmond Halley possessed sufficient generosity of spirit to recognize the significance of John Harrison's first sea-clock, or chronometer, in 1735, and had not Harrison been such a cautious perfectionist regarding his invention, Halley could well have greatly hastened Harrison's eventual receipt of the government prize. And with an astonishing absence of competitive spirit, George Graham, F.R.S., the leading clock and instrument maker in Europe, and one of the government's 'expert witnesses' in horological matters, openly expressed his admiration for Harrison's genius and the brilliance of design and execution displayed in his time-pieces. Indeed, Graham and other admirers of his fine sea-clock designs gave Harrison financial support out of their own pockets, Graham also making Harrison a significant interest-free loan to encourage him. Then in 1737, and after the rigorous testing of his first chronometer, 'H 1', the government made over an interim payment of £500 to Harrison. The same sum, in fact, that had been granted to Halley after 1720 for the wholesale re-equipping of the Royal Observatory, and one on which a 'middling gentleman' could keep his family in comfort for a whole year![29]

On 14 January 1742 (new style), and in his 86th year, Dr Edmond Halley died suddenly, probably from a heart attack, after, it was reported, downing a large glass of his favourite wine.

1.7 James Bradley

Edmond Halley died in office as Astronomer Royal. In the years before his death, however, he had clearly been grooming a man whom he hoped would be his successor. And very wisely, when Halley passed away, officialdom, for once, acted sensibly and appointed Halley's candidate. This was the 49-year-old Revd Dr James Bradley, F.R.S., sometime Fellow of Balliol College, Oxford, who had for the previous 21 years been Halley's co-Savilian Professor at Oxford. For as Halley held (and continued to hold until his death in 1742) the Savilian Chair of Geometry, so Bradley held the companion Savilian Professorship of Astronomy. Indeed, Bradley had been very much Halley's protégé, Halley having introduced the brilliant young Oxford astronomer to Newton, whose support he also enjoyed, and then backing him in his application for the Savilian Professorship in 1721 and for the office of Astronomer Royal. Bradley was the son of a modest country gentleman of Sherborne, Gloucestershire, and was sent to the local Northleach Grammar School and then to Balliol, and thereafter went on to an illustrious career.

Even while Halley had been in office, it is clear that James Bradley was a fairly regular visitor to Greenwich, in addition to the two men meeting in Oxford when discharging professorial duties, and at the Royal Society. We know, for instance, that in 1734 Bradley spent some time at Greenwich and assisted in the re-adjustment of some of the instruments, for as Halley and everyone else who had anything to do with astronomy knew, Dr Bradley was the most thorough and painstaking practical astronomer of the age.[30] Not only did he study angles in the heavens as a consummate celestial geometer, he also studied the behaviour of his instruments as engineering structures, being one of the first scientists to quantify – no doubt assisted by George Graham's 'pyrometer' researches – the seemingly invisible, 'microscopic' buckling that took place in an instrument in which brass and iron parts were joined together when it underwent a change in temperature. For brass expands more than iron when the temperature changes, and can produce errors of a few arc-seconds in an 8 ft radius quadrant.

Bradley was already famous across Europe when he became Astronomer Royal, his scientific 'immortality' having been assured in 1728, when he published a paper in the *Philosophical Transactions* announcing his discovery of the aberration of light. Using a type of telescope first devised by Robert Hooke in the 1660s, and also tried by Flamsteed at Greenwich, Bradley, like Hooke and Flamsteed before him, had attempted to measure the annual parallax of a particular star, *Gamma Draconis* (in the Constellation of Draco, the Dragon), which passes directly overhead in London. The telescope, called a 'zenith sector', consisted of a long tube that hung dead on pivots alongside its object glass (like a stationary pendulum), so that when an observer lay on his back and looked up the tube, he could only see a small patch of zenith sky. When *Gamma Draconis* transited the zenith meridian each day, its exact position could then be measured with a micrometer and timed with the precision regulator clock. And because no one knew precisely how much the earth's atmosphere bent starlight, it was thought safest to work with zenith stars because their light came down in a straight line with no bending or refraction.[31]

Bradley's zenith sector had a vertical telescope of 12½ ft focal length, adjusted with a precision micrometer measuring to a single arc-second. Yet when he came to analyse over a year's worth of observations of *Gamma Draconis*, he could detect no parallax motion. Yet what he did find was an entirely new and unexpected motion over a regular 6-month period at a *right angle* to the expected parallax. And while the zenith sector did not yield any figures from which he could compute the distance of *Gamma Draconis* (as we now know, even this instrument was insufficiently accurate to measure a stellar parallax), it dawned on Bradley that the displacement of the star image seen through the telescope must be caused by his moving *towards* the light of the star for 6 months and *away* from it for the next 6 months, resulting in an 'aberrated' or distorted image of the star's position as seen by an observer. And this in turn could only be so if the earth were moving in space about the sun.[32]

At last, two centuries after Copernicus wrote his heliocentric *De Revolutionibus*, the theory of the earth's motion had been demonstrated as a physical fact. Not only did Bradley's discovery blaze his name across scientific Europe, but the way in which he had made it set new standards in how to conduct cutting-edge research. For absolute and meticulous precision lay at the heart of it all. His 12½-foot zenith sector and micrometer, made by Graham, was probably the most accurate astronomical instrument in Europe in 1728, and four square in the British research tradition, it was Dr Bradley's private property. It was set up at the private house of one of Bradley's relatives, his astronomical uncle the Revd Dr James Pound, F.R.S., Rector of Wanstead, Essex, and visited by him at key points in the year, such as at the solstices and equinoxes. Then after 1748, by which time he been able to observe the moon through an 18-year cycle, he removed the zenith sector to the Royal Observatory, to be used as a zenith check upon the official instruments at the Observatory provided by the government. For by 1748, James Bradley had accumulated enough data with the zenith sector to announce yet another stunning discovery in celestial mechanics: the nutation. Taking its name from the Latin *nutare*, 'to nod', the nutation was a tiny yet measurable nodding of the earth's orbital system with relation to the moon.[33] And once again, it supplied a clear demonstration of Newtonian physics in action, as the varying elliptical orbit positions of the earth, moon, and sun produce slight but regular changes during the 18-year lunar cycle around the earth. Indeed, Bradley's discovery of both the aberration and the nutation constituted a spectacular synchronicity of the relationship between practical astronomy and gravitational theory. And at the heart of it all was James Bradley's acute awareness of the central role played by a progressive, critical instrumental accuracy, as each new discovery would suggest further problems that could only be solved after yet another level of instrumental accuracy had been achieved. For by the early to mid-eighteenth century, that creative dynamic operating between the scientist, the craftsman-scientist, the physical study of how materials behaved under changes of temperature or barometric pressure, and the design of a further, more advanced generation of precision instruments had evolved to a high degree. And by 1750, the Royal Observatory, Greenwich, was without doubt the main place in Europe where this dynamic was most conspicuously on display.

1.8 The English Craftsmen-Scientists

By 1700, and continuing well into the nineteenth century, London was to be the home of the most advanced scientific instruments in Europe, and it was not for nothing that in this burgeoning technological world the Royal Society, the wealth of the City of London, and the King's Observatory at Greenwich were neighbours both in spirit and in geography. For the political turmoils of the seventeenth century had made Great Britain unique in Europe in so many ways. At a time when France, Spain, Austria, Russia, and the Italian and German princely states had made absolute monarchy and a state-driven vertical patronage system the norm from Iberia to central Asia, Great Britain had come to develop a constitutional monarchy with limited political powers for the Sovereign, and all financial initiatives driven by Parliament. And while Great Britain undoubtedly had its poor, as shown in the engravings of William Hogarth and the mass open-air congregations of John Wesley, it also had the biggest and most financially autonomous middle class in world history. Eighteenth-century Britain saw a burgeoning of the professional classes, with successful business fortunes extending from those of great city merchants to rising shopkeepers (the source of the monies that backed Flamsteed, Halley, and Bradley), well-beneficed clergy, and legions of middling country gentleman, who acquired their money from land improvements and investment in shipping, East India Company stocks, and early Industrial Revolution projects amongst other things.

And as many of these people took an intelligent interest in science and were willing to pay handsomely for fine clocks, barometers, and astronomical instruments with which to fit out private observatories, and to acquire valuable libraries, there came into being an established market-driven demand for fine technology. What is more, the craftsmen who came to meet that demand were not looked down upon as mere 'mechanics', for the production of fine telescopic optics or exquisitely graduated mathematical scales presumed not only skill, but also a firm understanding of the technical problems that their wares were expected to solve. It is for this reason, therefore, that the upper échelons of the scientific instrument-making trade could become comfortably off and socially accepted as gentlemen in their own right, winning Admiralty awards, and in certain cases even Fellowships of the Royal Society, or receiving the Society's Copley Medal, the Nobel Prize of the eighteenth century.

Thomas Tompion, who made Flamsteed's early clocks and instruments, retired to Bath on the profits of his watch and clock business, presenting the city with a fine long-case clock which to this day still marks the hours in the Pump Room. His successor, George Graham F.R.S., even sat on the Royal Society Council and advised the government about the longitude. Both Tompion and Graham were buried in Westminster Abbey, an honour granted to Newton, but not to any Astronomer Royal. John Harrison, the marine chronometer inventor, won the Copley Medal, backed by a magnificent citation from the President of the Royal Society in 1749, while his son and assistant, William Harrison, became an F.R.S. and a member of the Georgian scientific élite in his own right.

John Bird, who re-equipped the Royal Observatory for Bradley after 1750, was not only financially successful and left a handsome fortune, but became in his generation scientific Europe's first port of call on matters of precision, and had many Fellows of the Royal Society amongst his personal friends. Then from the mid-eighteenth century and extending into the nineteenth, there was John Dollond F.R.S. and his optical dynasty, then Dollond's son-in-law, the eminent Jesse Ramsden F.R.S., Edward Troughton F.R.S., and William Simms F.R.S. Even the cosmologist, Sir William Herschel, F.R.S., L.L.D. and Knight Guelph, was part of the same movement.

For Herschel had begun life in Germany as a humble oboist in the Hanoverian Foot Guards Band who, once in England and working as a musical impresario in Bath, discovered his own genius for making exquisite telescope mirrors, and made several thousand pounds from the manufacturer of fine reflecting telescopes, which success story he crowned by marrying the young widow of a City of London merchant. Herschel's combination of manual dexterity, intellect, and obvious personal charm soon saw him elected to the Royal Society and cementing firm friendships with Sir Joseph Banks, President of the Royal Society, the Revd Dr Nevil Maskelyne, F.R.S., Astronomer Royal, and even King George III himself.[34] Only in England, with its open élite of self-made money and intellect, would a career like Herschel's have been possible, and it was not for nothing that when visiting French *savants* attended Royal Society meetings, they sometimes expressed surprise at being sat next to a working clock-maker or lens-maker on terms of social equality, for back in Paris such men would have been ranked as little more than servants.

Indeed, so appealing could the profession of scientific instrument-maker be by the mid-eighteenth century, that some men migrated into it from other backgrounds. James Short, for instance, became an M.A. of St. Andrews University, and while qualifying to serve as a Minister of the Church of Scotland at Edinburgh University found that his self-taught skills as an Edinburgh telescope-maker brought in good money and friendships in the University's scientific establishment, especially with the famous astronomer Professor Colin Maclaurin. Then around 1739, James Short moved to London as a telescope-maker. Not only did he make a handsome fortune of £20,000 in the process, but he was elected an F.R.S., sat on the Society's Governing Council, advised on the best way to observe the transit of Venus of 1761, and was a major – albeit unsuccessful – candidate for the post of Astronomer Royal after Bliss's death in 1764. He was also a friend and generous supporter of John Harrison, as well as of other innovators.[35]

One has only to look at the exquisitely finished portraits of instrument makers such as Tompion, Graham, Bird, or Ramsden to grasp their status and firm social assurance as, looking confidently out from under their periwigs, their faces could easily be imagined to belong to a Bishop, a Judge, or a Cabinet Minister. For they had all become gentlemen. All of these above-mentioned scientists-craftsmen moreover, and a good few besides, had close connections with the Royal Observatory, spanning the period from 1676 to 1850. They either designed and built new large, capital instruments for Greenwich, were brought in to conduct tests on instruments, worked closely with the Astronomer Royal on a particular physical investigation, or else advised, and sometimes even sat upon, the Admiralty's scientific Board of Longitude after its formation in 1714. And the scientist-craftsman who worked most closely with James Bradley was John Bird.

1.9 James Bradley's Instruments and the Problems of Heat

A clear indication of how rapidly not only the science of astronomy was moving by the mid eighteenth century, but also the 'science of materials', is found in the growing awareness by 1750 that the Royal Observatory's instruments were no longer adequate. George Graham's magnificent 8 ft radius quadrant built for Halley in 1725, for instance, was found to have become distorted by a crucial hair's breadth amount over less than a quarter of a century, giving it an error of 16 arc-seconds.[36] Of course, this had nothing to do with any faults in George Graham's exquisite workmanship, but emerged from the growing realization that a quadrant with an engraved brass limb riveted to a wrought-iron frame would produce stress distortions due to the fact that, whenever the temperature changed, the brass would expand more than the iron. And as a result, Graham's beautifully engraved scales would develop inequalities.

In some ways it is ironic that this should have happened with a Graham quadrant, for George Graham had been one of the first scientists to conduct experimental investigations into the thermal properties of metals, although the purpose of these studies had been to devise a self-compensating pendulum for clocks rather than for quadrant scales.

Indeed, in 1726 Graham had devised his mercury bob clock pendulum: a glass cylinder containing mercury now replaced the conventional brass ball on the end of the pendulum. By determining exactly the expansion coefficients of iron (used for the pendulum rod) and mercury, Graham had derived a formula by which, through any given temperature rise, the iron pendulum rod would expand downwards (and hence slow down the clock), but this would be countered by the mercury in the glass cylinder bob expanding upwards. The idea behind the invention was to maintain automatically the same length of oscillation, between the pendulum's suspension point, and the mass centre of the mercury bob. (I would suggest that this was perhaps one of the world's first 'smart' inventions, in so far as it was self-compensating, responded to environmental changes automatically, and required no human interference.). And shortly afterwards, John Harrison invented an alternative temperature compensation device, in his 'gridiron' pendulum, in which the *downwards* expansion of five steel rods was offset by the *upwards* expansion of four brass rods. Both pendulum designs would become pivotal to the eighteenth-century development of the precision astronomical regulator clock.[37]

Of course, Graham's clocks for Halley came to be equipped with such self-compensating bobs (even though his pre-1726 ones did not have them at first), and still, some 280 years later, they keep amazingly accurate time. In 1725, however, Graham clearly had not reckoned the metal expansion problem to be of sufficient concern as to take note of it in his brass and iron 8-foot quadrant. But by 1746, this problem had indeed become apparent – and had even been quantified – simply because Graham's quadrant scales had been the most perfect ever drawn and hence made the tiny thermal error obvious, as would not have been the case in Flamsteed's less precise 7-foot Mural Arch or sextant of 70 years before. But as George Graham was now elderly (and would die in 1751), Dr Bradley sought a solution from the hands of his commercial and scientific successor: John Bird.

Having secured a government grant of £1,000 to bring the instrumentation of the Royal Observatory up to date once again, John Bird was the obvious person for James Bradley to approach. For Bird had developed the science of practical geometry for the production of exact mathematical scales to a new point of excellence, giving extra seconds of arc reliability on top of even the best of George Graham's work. Bird achieved this from a complex, cross-checking geometry, performed with precision beam-compasses and a meticulously subdivided linear 'scale of equal parts' of his own devising. The new quadrant upon which he lavished these skills seemed, at first glance, identical to Graham's 1725 instrument: 8 ft in radius, with a similar system of support girders, precision vernier and micrometer scales, and the pair of 90° and 96-part scales that could be cross-checked against each other. But where the old and the new quadrants differed most fundamentally was in regard to their thermal homogeneity, for the new Bird quadrant consisted of one single metal for every part of its construction: the highest quality brass. And as everything – scales, frame, rivets, screws, and so on – were of the same material, they would expand and contract in mutual synchronicity, so that the geometrical unity of the whole remained unchanged. It was a brilliant design concept. And by the time that the quadrant was fully in operation by 1753, it was by far the most accurate astronomical instrument ever made, and as Bradley discovered after meticulous testing, it could measure a vertical, or declination, angle in the meridian to plus or minus one single arc-second. The quadrant consumed £300 of the £1,000 grant.[38]

Of course, George Graham's magnificent original and inspirational quadrant still remained in use, but now it was turned through 180 degrees and re-hung on its heavy stone supporting wall so that it measured angles in the northern meridian, to monitor the angle of the Pole Star, and the transits of Capella and other circumpolar stars. This was still important work, but less so than working in the southern meridian, for the sun, moon, and planets were never seen in the north. A 'retirement' job for a venerable masterpiece, in fact.[39]

This configuration of instruments, with the new Bird quadrant on the southern meridian and the older Graham instrument on the polar or northern one, which now covered the entire meridian arc from horizon to horizon through 180 degrees, was to be used not only by Bradley, but also by his two successors in the office of Astronomer Royal, the Revd Nathaniel Bliss and the Revd Dr Nevil Maskelyne. Only after John Pond succeeded Maskelyne in 1811 was it felt necessary to retire both 8-foot quadrants, for by then the quadrant had been upstaged in accuracy by the new Meridian Circle. But in 1753 Bradley had a set of instruments which made Greenwich the accepted international benchmark of accuracy as far as angle measurement was concerned.

Yet Bradley's basic method of working was really the same as Flamsteed's and Halley's: objects would be observed, timed, and measured as they came to the meridian, and the resulting data analysed, reduced, and published as astronomical tables. Where Bradley was so significant, however, was in his new standards of accuracy: a significance, moreover, which derived not just from a new quadrant, but from a whole new system of instrument error cross-checking which Bradley built up around it. For one thing, in addition to the new quadrant and some other pieces, Bradley's government grant had been used to commission from John Bird a much larger, more powerful, and more easily checked transit instrument which replaced the one which Graham had built for Halley in 1721. Bradley used this new, refined transit instrument to constantly monitor the exact vertical plane of his quadrants. If, for instance, Bradley and his assistant observed the simultaneous meridian transit passage of a star, with the transit instrument and the quadrant respectively, and there was a discrepancy of timing between the two for a given zone on the quadrant's plane, then it became obvious that the plane was not perfectly vertical and needed to be corrected. Likewise, Bradley used his zenith sector as a vertical check, to ensure that both his Bird and his Graham quadrants were perfectly adjusted to the zenith.[40]

Central to James Bradley's achievement as an astronomer was his recognition of the need to track down and compensate for every possible source of error when making an observation, and he did this by surrounding his entire observing procedure with a network of inter-instrument cross-checks. He made the transit instrument check the clocks, the transit instrument and zenith sector check the quadrants, and the whole system to check itself, root out specific errors, and correct them. Then, in addition to the purely mechanical accuracy of his instruments, Bradley came to realize that a hitherto unsuspected source of error must be monitored as well: the state of the earth's atmosphere at the time of the making of an observation. For Bradley came to realize that barometric pressure, temperature, and even moisture content affected the refractive properties of the air through which a measured ray of starlight was passing, and after 1750 the Royal Observatory began to make a systematic record, with barometer and thermometer, of the daily state of the air. Bradley's experiments on the changing optical properties of the atmosphere made possible the construction of standard tables, so that an instrumental observation could be brought as close to perfection as feasible.

In his great analytical work in astronomy, the *Fundamenta Astronomia* (1818), the eminent German astronomer Friedrich Wilhelm Bessel acknowledged that the meticulous accuracy of Bradley's observations made them the earliest that could be absolutely relied upon when conducting long-term analyses of the gravitational and other behaviour of astronomical bodies.[41] For Bradley's well-publicized research methods were to be adopted as standard practice in all observatories worldwide, whenever positional measurements of the highest order were to be made. And they remained in use – with obvious refinements – so long as visual 'hand and eye' meridian astronomy was performed. For only in the late twentieth century, when computer-controlled and photographic methods came to replace the older manual ones, did these procedures fall out of use in front-rank research.

1.10 The Longitude by Lunars – At Last

Over 80 years after the establishment of the Royal Observatory, Greenwich, a variety of factors coalesced to make the lunars method of finding the longitude at sea practicable at last. Of course, no one in 1675 could have dreamt that it would have taken so long. But that is in the nature of scientific research, especially when the pursuit of a practical goal must be preceded by a vast amount of pure research. For as later generations of scientists across the entire sweep of the sciences have discovered, problems are never predictable in advance. Newton's fundamental analyses of the lunar orbit, conducted within the context of gravitational theory, had provided the valuable intellectual instrument with which to address the longitude. But before theory could become practice, it was necessary to build up a formidable harvest of primary observational data on which the calculations could be based, and, as we have seen, this depended on a succession of improvements in angle-measuring technology. Through three generations of it indeed: from Tompion and Abraham Sharp, who made Flamsteed's instruments, to Graham's instruments for Halley, and on to the quadrants made by John Bird for Bradley and the other astronomers who used not only Bird's instruments but also Bradley's methods. It is true that the improvement in accuracy between 1689 and 1750 had only been from an error of 10–12 arc-seconds at best to one of 1 or 2 arc-seconds, but that improvement turned out to be crucial.[42]

Yet the man who produced the first set of lunar tables which showed that the longitude by lunars was within reach at last was not Bradley. It was a German, Tobias Mayer, of the Göttingen Observatory who also happened to be, due to the Hanoverian dynastic connection, a subject of Great Britain's King George II who was also the Elector of Hanover. And Mayer was fortunate to be able to work at Göttingen with an excellent Bird quadrant of 6 ft radius – one of several fine instruments which Bird was commissioned to build for foreign observatories. And when their results were compared, it turned out that Bradley's and Mayer's observations agreed to within 2 arc-seconds. Tobias Mayer's tables were examined by the British Admiralty in 1755 and after rigorous testing later secured a £3,000 award for Mayer's widow from the Board of Longitude under the Act of 1714.[43]

It was five years after James Bradley's death in 1762 that his successor-but-one, the Revd Dr Nevil Maskelyne, was able to use the Greenwich and Mayer's observational data to produce the first *Nautical Almanac* in 1767. For here at last were lunar tables, and instructions for their use, by which a sailor could employ the moon and stars to fix a ship's position pretty well anywhere on the earth's surface. And one of their first, and triumphal, uses was on Captain James Cook's first voyage of discovery to the 'Great South Sea' in 1768, which not only began the serious scientific cartography of the Pacific Ocean, but also led to the discovery of New Zealand and eastern Australia.

What is often forgotten, however, when discussing the eventual success of the lunars method of longitude finding, is the shipboard navigator's ability to measure the angle between the moon and a given star to within a minute or less from the heaving quarterdeck of a vessel on the high seas. The invention of John Hadley's reflecting Mariner's Octant or quadrant in 1731 had been a major step in that direction, but octants with a radius of 12 or 18 inches, held close to the navigator's body when in use to maximize stability, were not sufficiently accurate to enable lunar angles to be reliably taken.[44] In the 1750s, however, Bird engraved a Hadley Octant and a 'reflecting circle' of larger angular amplitude using his exquisite dividing methods, and test observations suggested that such instruments could be used, in conjunction with high-quality, observatory-generated lunar tables, to find the longitude at sea. But reflecting circles and octants such as Bird's were effectively one-offs, and very expensive, just like John Harrison's beautiful chronometers, and before the lunars method could be routinely used at sea, the navigator needed not only his reliable volume of tables, but also an equally reliable instrument with which to shoot lunar and stellar angles from the deck of his ship.

This led to what one might think of as an early industrialization of practical astronomy and navigation, as first Jesse Ramsden F.R.S., and then Edward Troughton F.R.S., developed a succession of scale-dividing engines after 1766, whereby an accurately divided master-circle, 3 or 4 ft in diameter, could be used to execute a reduction division, of equal accuracy to the master-circle, onto the bodies of brass octants.[45] And because it was often convenient for a shipboard navigator to shoot lunar and stellar angles of more than 90° on a given night, Ramsden and others developed the sextant, which gave a reliable angular amplitude of up to 120 degrees of arc. So by the time that the young Horatio Nelson was learning navigation as a midshipman in the early 1770s, the lunars method had become not only wholly practicable, but also relatively cheap. For while that much more widely publicized longitude-finding instrument, the John Harrison chronometer, cost around £500 apiece to produce, the lunar method was cheap and commonplace, requiring no more than a mass-produced volume of tables – the *Nautical Almanac* in 1767 and thereafter – costing a few shillings, and a 'mass-produced', machine-graduated sextant costing around ten guineas.[46]

After nearly a century, the nation's investment in the King's Observatory in Greenwich Park had fulfilled the intentions of its founders and made it routinely possible to use the moon to find the longitude at sea.

1.11 Nathaniel Bliss, Astronomer Royal 1762–1764

The Reverend Nathaniel Bliss, F.R.S., was 62 years old when he succeeded his friend and to some degree patron James Bradley, and perhaps it was hoped that following his appointment he would be like Halley and give 20 years of sound and active service. One presumes that he must have been active and vigorous at 62. Yet sadly, he died after having held the office of Astronomer Royal for only two years (Figure 1.4).[47]

Like Bradley before him, Nathaniel Bliss was the son of a modest country gentlemen of Gloucestershire, and on going up to Pembroke College, Oxford, he clearly entered that Oxford astronomical world that revolved around Edmond Halley. Bliss was a meticulous practical observer, skilled in the use of astronomical instruments,

FIGURE 1.4 Nathaniel Bliss succeeded Bradley as Astronomer Royal. (Reproduced from E. Walter Maunder, *The Royal Observatory Greenwich: A Glance at Its History and Work* (London: The Religious Tract Society, 1900).)

and like Halley, Bradley, and the young Thomas Hornsby, was on excellent terms with those two astronomical aristocrat-lawyers, Thomas and George Parker, father and son, of Shirburn Castle, Oxfordshire, and both Fellows of the Royal Society, better known by their titles, the First and Second Earl of Macclesfield. When Halley died, holding the joint offices of Astronomer Royal and Savilian Professor of Geometry at Oxford, and Bradley, already Savilian Professor of Astronomy, moved to Greenwich, Bliss succeeded his mentor as Geometry Professor. Then when Bradley died, Bliss in turn moved to Greenwich, while still retaining his Oxford Chair.

At Greenwich, as Astronomer Royal, Nathaniel Bliss inherited not only Bradley's superb Graham and Bird quadrants and other instruments, but also Bradley's method of observation, being based, as we have seen, on meticulous procedures of cross-checking. He inherited in addition Bradley's old assistant, Charles Green, himself a meticulous worker, who in 1768 would sail with Captain Cook on *H.M.S. Endeavour* as the official astronomer for the observation, from Tahiti, of the Venus transit of 3 June 1769. A voyage, alas, which would claim his life as a result of the 'bloody flux', or acute dysentery. For Green was one of 24 men whom Cook described as dying from this disease in January 1771, as the ship was homeward bound from Batavia in the East Indies, a notoriously fever-ridden place.[48]

And just as Bradley had worked on special projects at Greenwich with Halley in the 1730s, so Nathaniel Bliss often joined his fellow Savilian Professor friend, Bradley, to make observations after 1742. It is a pity, however, that his life was cut short after only 2 years as Astronomer Royal, for Bliss had no time to become involved in any new research project or to make his mark on the Royal Observatory. Indeed, his only significant achievement while in office was the observation of an annular eclipse of the sun, which took place on 1 April 1764 and which he reported in the Royal Society's *Philosophical Transactions*.[49]

Following the death of Nathaniel Bliss, the Royal Observatory's line of Oxford Astronomers Royal came to an end. For from the Revd Dr Nevil Maskelyne onwards, who was appointed in 1765, the Directorship of the Royal Observatory was to remain a Cambridge preserve down to the end of the Observatory's existence in the late twentieth century. Yet ironically, the man who succeeded to Bradley's Savilian Professorship of Astronomy at Oxford in 1762 was to be the driving force behind the establishment of a new state-of-the-art observatory in Oxford. This was the Revd Dr Thomas Hornsby, F.R.S., and the observatory which he was to direct between 1772 and 1810 was that paid for by the University's munificent patron, the Radcliffe Trustees. Equipped with a magnificent pair of 8 ft radius brass quadrants, a transit, zenith sector, equatorial instrument, and other pieces by John Bird at a cost of £1,300 (it was Bird's last major commission before he died in 1776), the Radcliffe Observatory was reckoned by the visiting Danish astronomer, Thomas Bugge, in 1777 as one of the finest and best-equipped in Europe. But undoubtedly Greenwich was its prototype.[50]

1.12 Conclusion

It is hard to overestimate the role played by the Royal Observatory in the development of British and wider European astronomy after 1675. For as Flamsteed traced out in his 'Prolegomena', it emerged as the natural outcome of a rich tradition of observational astronomy, with its roots in the European observatories of Hevelius, Tycho, and Walther, combined with the vigorous mathematical and inventive traditions of late Tudor and Stuart England. Yet several factors conspired to give Greenwich a unique and powerful importance. For very significantly, by 1675 those crucial instrumental innovations, the telescopic sight, the screw micrometer for reliably delineating very small angles, and the pendulum clock, had just passed through their experimental stages, and were becoming 'established' technologies. But it was during Flamsteed's 44-year reign at the Royal Observatory that they were rigorously put through their paces on a nightly basis, transformed from ingenious novelties into the established devices of precision measurement, and built into a system of practical observation which by the early eighteenth century was being followed in every serious observatory in Europe. And the basic devices and procedures of usage remained, albeit with constant improvement and refinement, as the standard ways to measure right ascension and declination angles wherever western astronomy was practised across the globe, until even more accurate electronic measuring techniques replaced manual ones in the late twentieth century.

And in spite of its chronic under-funding and eccentric and often erratic instrument provision in the early decades, and dependence on the goodwill of its independently well-off first Astronomers Royal, the institution's status as a Royal Observatory, with close connections to the Ordnance Office, Admiralty, and Board of Longitude, gave it a vital power of endurance. For neither Tycho's nor Hevelius's observatories operated for more than 35 years, and while King Louis XIV of France had established an observatory just outside Paris after 1667, which in its early days had served as a work-place for such astronomical luminaries as Jean Picard and Giovanni Domenico Cassini, the quality of its instrumentation lagged conspicuously behind that of Greenwich by 1700, as it came increasingly to fulfil the role of scientific showpiece of an absolutist state rather than that of a dynamic research institution. But Greenwich's endurance, its succession of gifted and determined Astronomers Royal, grants for new instruments after 1720, and the elusive carrot of finding the longitude by lunars, established the King's Observatory in Greenwich Park as the premier observatory in Europe, especially in the sphere of celestial cartographic astronomy.

2

Greenwich, Nevil Maskelyne and the Solution to the Longitude Problem

Mary Croarken

CONTENTS

2.1 Introduction

On 16 March 1765 Nevil Maskelyne (1732–1811), the recently appointed fifth Astronomer Royal, took up residence at the Royal Observatory, Greenwich, where he lived and worked for almost 46 years until his death in January 1811. While Maskelyne was involved in many mathematical and scientific projects through the Royal Society, he is best known for the creation of the annually published *Nautical Almanac*, a publication which provided a cheap and accessible method of finding longitude at sea, thereby improving the safety of navigation at sea (Figure 2.1).

Like his predecessors as Astronomer Royal from Flamsteed onwards, Maskelyne was specifically appointed to conduct astronomical work with the practical outcome of improving navigation at sea. The royal warrant confirming his appointment on 8 February 1765 instructed Maskelyne

> forthwith to apply yourself with the most exact Care and Diligence to the rectifying the Tables of the Motions of the Heavens, and the Places of the fixed Stars, in order to find out the so much desired Longitude at Sea, for perfecting the Art of Navigation.[1]

FIGURE 2.1   Nevil Maskelyne, Astronomer Royal 1765–1811. (Royal Museums Greenwich ROG 11498 © National Maritime Museum, Greenwich, London.)

Maskelyne followed the terms of his appointment to the letter, and while at Greenwich he was instrumental in the controversies surrounding the solution to the Longitude Problem, in making the solution accessible to all seamen and in laying the foundations for Greenwich being designated as the internationally recognised Prime Meridian in 1884 and the baseline for the world's timekeeping system.

2.2 The Longitude Problem

By the mid-eighteenth century sailors could readily find their latitude by observing the altitude of the sun, but determining longitude at sea with any accuracy was not possible. Longitude on land could be accurately found astronomically by observing the eclipses of the moons of Jupiter[2], but finding longitude at sea, however, was a much more difficult proposition – the motion of a ship making the required accuracy of observation impossible. This inability to accurately determine a ship's exact position could have disastrous consequences for human life as the wreck of

Sir Cloudesley Shovell's fleet off the Scilly Isles in 1707 with the loss of 2,000 men demonstrated.[3] By the early eighteenth century it had become obvious that British military, imperial and trade interests were being hampered by the lack of a reliable method of finding longitude at sea. In 1714 the British Government offered a reward of £20,000 (a sum worth over £2m today) for a practical solution to the Longitude Problem and created the Board of Longitude to oversee the administration of the prize. The terms of the prize were specified in the Longitude Act and stated that £10,000 would be paid out for a practical method of finding longitude at sea to an accuracy of 1° or 60 nautical miles, and £20,000 if found to an accuracy of half a degree or 30 nautical miles. The method was to be tested on a sea voyage to the West Indies. The Longitude Act also allowed the Board of Longitude to pay out smaller sums for promising experimental work.[4]

The theoretical principles of finding longitude were already well known and were intimately connected with time. The difference in longitude between two places on the surface of the earth is equal to the difference in local time of the same two places. Longitude is expressed as degrees east or west of a specified meridian, and today, thanks to Maskelyne's work, Greenwich is internationally regarded at the standard or prime meridian. Finding longitude at sea is the same problem as knowing what time it is locally and what time it is at Greenwich. Local time can be found by observing the sun, but knowing the time at Greenwich at the same instant was a bigger problem.

One obvious way to find longitude is therefore for a navigator to 'carry' Greenwich time using an accurate clock and to compare Greenwich time as read from the clock, or chronometer, with local time as observed. A one-hour difference in time is equivalent to 15 degrees longitude; four minutes difference in time is equivalent to 1 degree longitude. However in the early eighteenth century the technology did not exist to make chronometers that could keep sufficiently accurate time under the extreme conditions found on board an eighteenth century ship as it travelled the world for many months or years. Not only did the motion of the ship affect the going of chronometers but changes in temperature and humidity affected the physical properties of the materials used to construct them and hence affected the predictability of their accuracy.

Alternately the movement of the moon against the background of the fixed stars could be used as a clock with which to calculate longitude – the so-called lunar distance method. The movement of the moon is theoretically predictable and so, therefore, is the distance of the moon from a fixed star, i.e. the lunar distance, at any given instance. If a navigator knows the time at Greenwich at which a specific lunar distance will occur, he can calculate his longitude as the difference between Greenwich time and local time when the observed lunar distance occurred.[5] The lunar distance method had long been known in theory,[6] but a method of predicting the place of the moon with sufficient accuracy was not available in the early eighteenth century.

The Longitude Problem was eventually solved in the 1760s using both the chronometer and the lunar distance methods. While John Harrison was awarded the prize in 1773 for his watch H4, it was Nevil Maskelyne who provided a solution that was more rapidly made available to sailors and had a more immediate effect on the accuracy and safety of navigation.

2.3 John Harrison's Watch

The story of how John Harrison (1693–1776) eventually won the Longitude Prize in 1770 for his watch H4 is a well-known one often told.[7] In popular myth, based largely on Harrison's own propaganda[8] and the intervention of King George III, Maskelyne is portrayed as the villain of the piece and is accused of trying to prevent Harrison winning the prize by promoting his own method of lunar distances by not giving Harrison's watch the credit it deserved. Yet Howse, Maskelyne's biographer, states that this reputation

> was certainly not one that was held generally in his own day, nor is it in any way justified by today's research. He was a member of the Board of Longitude, appointed by Parliament to advise in the award of large sums of public money. There is no evidence whatsoever that Maskelyne at any time abused his position as a public servant in order to further his own ends, still less to line his own pocket.[9]

Harrison, a Yorkshire carpenter, began working on clocks intended to be accurate enough to determine longitude at sea in the 1730s. His experiments were promising, and he was encouraged, both financially and intellectually, by Edmund Halley (the then Astronomer Royal), George Graham (the most influential clockmaker of the period), the Royal Society (being awarded their Copley Medal in 1749) and the Board of Longitude who awarded Harrison sums of money from time to time to support the work. In 1761 Harrison's H4 watch was ready for sea trials, and the watch was tested on a voyage to Jamaica. The results were impressive, but the Board of Longitude wanted further proof that the watch was accurate and reliable and a further trial to Barbados was arranged in 1764 with Maskelyne as part of the team testing the watch (Figure 2.2).

The results of the trial demonstrated that Harrison's watch H4 could indeed be used to determine Longitude at sea as specified in the Longitude Act, but still the Board of Longitude wanted further assurance that the method could be replicated. In 1765 the Board of Longitude awarded Harrison £10,000 (minus monies he had already received for the development of the watch) if he promised to disclose how the watch was made and promised a further £10,000 when a replica watch could be built to a similar accuracy. By now disillusioned and disappointed that a lifetime's work was not being rewarded as he felt he was entitled, and convinced that Maskelyne was working against him, Harrison reluctantly disclosed how the watch had been made and had to wait while another watchmaker, Larcum Kendall, copied it. Kendall's copy, K1, was tested during Captain James Cook's second voyage 1772–1775, and it proved beyond doubt that the watch could be successfully applied to finding longitude at sea to the required accuracy. In the meantime, Harrison (now almost 80 years old) made a personal appeal to King George III who acted swiftly to ensure that Harrison was awarded the remainder of the prize in 1773.

The Longitude Problem was solved, but the price of making watches to Harrison's standards was high. Kendall's copy had cost £200, and it would be over 50 years before the cost of such watches made them commonplace on all British ships. In the interim Nevil Maskelyne, based at the Royal Observatory, Greenwich, provided an alternative solution which won no prizes but was almost universally employed.

FIGURE 2.2 John Harrison's watch H4 – winner of the Longitude Prize. (Royal Museums Greenwich ROG D0789-1 © National Maritime Museum, Greenwich, London.)

2.4 Making the Lunar Distance Method Practical

In 1761 Maskelyne, then an up-and-coming young astronomer, was sent by the Board of Longitude to observe the transit of Venus on the island of St Helena in the southern Atlantic Ocean as part of a Royal Society effort to use the transit to determine the distance of the earth from the sun. On the journey Maskelyne was also to test a set of lunar tables sent to the Board of Longitude by Tobais Mayer (1723–1762) for their suitability in finding longitude using the lunar distance method.[10] Maskelyne was very impressed with both the accuracy of Mayer's tables and the lunar distance method generally. On his return to England in 1763 Maskelyne published *The British Mariner's Guide*[11] in which he described the necessary observations and calculations and provided extracts from Mayer's tables as well as worked examples from his voyage (Figure 2.3). The one drawback of the method was that after the observations to determine lunar distance and local time had been taken, the resulting calculations took four hours to complete and required the use of a book of logarithms. This was unrealistic in the setting of most eighteenth century ships. However Maskelyne realised that a large part of the calculations focused on finding the predicted lunar distances at Greenwich and did

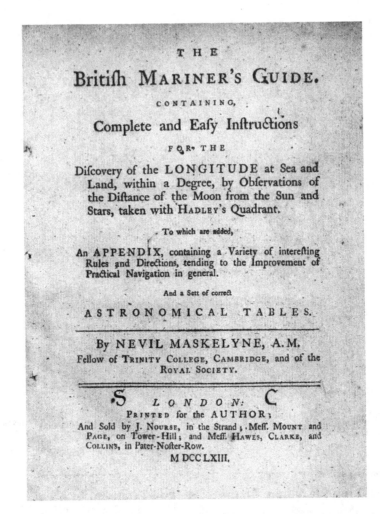

THE

Britifh MARINER'S GUIDE.

CONTAINING,

Complete and Eafy Inftructions

FOR THE

Difcovery of the LONGITUDE at Sea and
Land, within a Degree, by Obfervations of
the Diftance of the Moon from the Sun and
Stars, taken with HADLEY's Quadrant.

To which are added,

An APPENDIX, containing a Variety of interefting
Rules and Directions, tending to the Improvement of
Practical Navigation in general.

And a Sett of correct

ASTRONOMICAL TABLES.

By NEVIL MASKELYNE, A.M.
Fellow of TRINITY COLLEGE, CAMBRIDGE, and of the
ROYAL SOCIETY.

S *LONDON:* C
PRINTED for the AUTHOR;
And Sold by J. NOURSE, in the Strand; Meff. MOUNT and
PAGE, on Tower-Hill; and Meff. HAWES, CLARKE, and
COLLINS, in Pater-Nofter-Row.
M DCC LXIII.

FIGURE 2.3 Title page from Maskelyne's British Mariner's Guide, 1763. (Royal Museums Greenwich ROG E8872 © National Maritime Museum, Greenwich, London.)

not depend on the observations made at sea. In theory the predicted lunar distances could therefore be pre-computed on land and provided to seamen to use during their calculations.[12] This would have the effect of reducing the calculation to approximately half an hour and was, therefore, a much more realistic proposition at sea. The concept of pre-computing the lunar distances was the fundamental driving force behind Maskelyne's creation of the *Nautical Almanac*.

In 1763 the Board of Longitude sent Maskelyne to Barbados to test both Harrison's watch and an improved set of lunar tables sent by Mayer to the Board of Longitude shortly before his death in 1762.[13] Harrison perceived Maskelyne to be a champion of the lunar distance method and objected to his testing both methods although as discussed above, there is no evidence to suggest that Maskelyne did not carry out the task fairly. During the test Maskelyne found both the chronometer method using Harrison's watch and the lunar distance method to be successful. On his return from Barbados Maskelyne was appointed on 8 February 1765 as Astronomer Royal in succession to Nathaniel Bliss, and his first task was to report to the Board of Longitude meeting on 9 February on the trials of both Harrison's watch and Mayer's lunar tables.

As a result of the favourable reports the Board received from Maskelyne and others, the Board of Longitude decided to conditionally award Harrison the main prize (see above), Mayer's widow £3,000 for the lunar tables, and £300 to Leonhard Euler on whose lunar theories Mayer's tables were based. At the same meeting, Maskelyne suggested that annually producing an almanac containing pre-computed lunar distances along with all the astronomical data required for navigation was the most effective way of enabling the lunar distance method of finding longitude at sea. The Board agreed and charged Maskelyne, in his capacity as Astronomer Royal, with the task of producing two editions of what became known as the *Nautical Almanac* to demonstrate both the usefulness of the idea and Maskelyne's ability to prepare the publication.[14]

2.5 Plans for the *Nautical Almanac*

Maskelyne immediately began work on the *Nautical Almanac*, the Board of Longitude having given him less than two years to plan, calculate and publish the first edition for 1767. In many ways the *Nautical Almanac* was similar to other almanacs then available, such as the French *Connaissance des Temps*. In 12 pages per calendar month, the *Nautical Almanac* provided calendar dates and predicted positions of the sun, the planets and the fixed stars, but its novel feature was that its predictions of the position of the moon were based on Mayer's tables which were more accurate than any that had gone before and, crucially for finding longitude, it provided pre-computed lunar distances at regular intervals for every day of the year. It also provided calendar data, tables giving the position of the sun, the eclipses of Jupiter's satellites, the position of the planets and the position of the moon.

Maskelyne was now faced with the problem of setting up a system of annually producing all these tables in a timely, cost-effective and, most important of all, accurate manner. In the *British Mariner's Guide* Maskelyne had already begun to refine the steps necessary to calculate the individual table entries, but here the problem was to efficiently repeat the calculation for every 3 hours of every day of the year. Maskelyne's solution was to begin to develop what was to become a network of human computers. Initially, in June 1765, Maskelyne hired five men to undertake the work under his close supervision. Two, Israel Lyons (1739–1775) and George Witchell (1728–1785), were tasked with computing the 1767 *Nautical Almanac*, and another two, John Mapson (fl.1765–1771) and William Wales (1734–1798), were allocated the 1768 edition. Richard Dunthorne (1711–1775) was appointed in July to ensure that the tables compiled were error free and to correct the proofs of the printed tables. Work proceeded more slowly than Maskelyne had hoped, and in July 1766 Mapson and Wales were instructed to help out with the 1767 edition (Figures 2.4 and 2.5).[15]

The first edition of the *Nautical Almanac* was for the year 1767 and was published in January that same year, the 1768 edition following soon afterwards. These first two editions of the *Nautical Almanac* acted as a proof of concept for the lunar distance

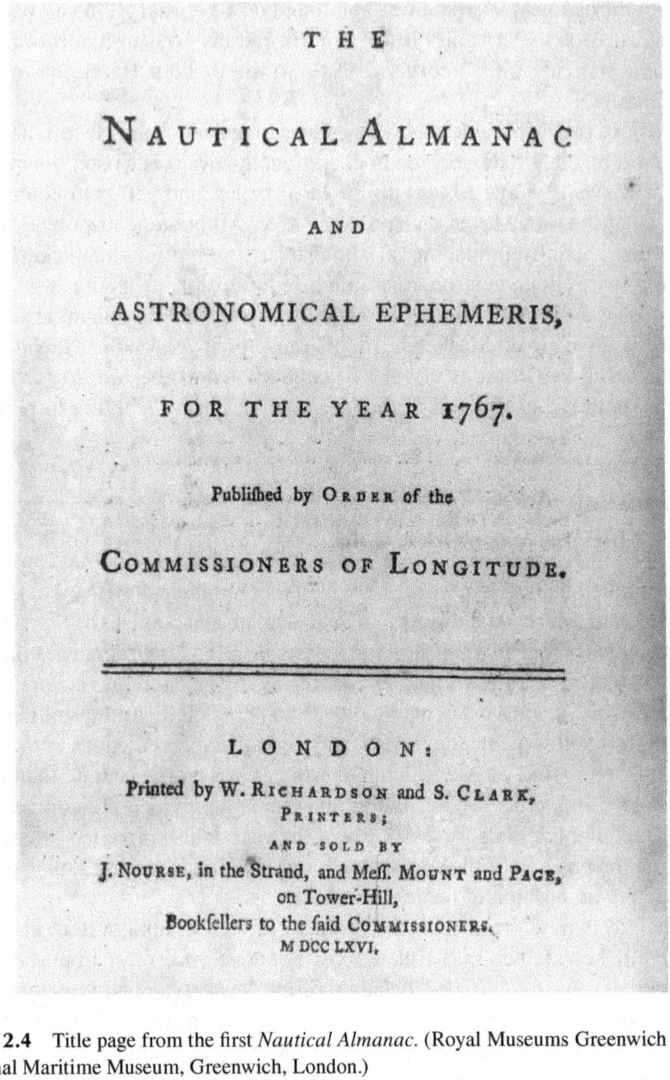

THE

NAUTICAL ALMANAC

AND

ASTRONOMICAL EPHEMERIS,

FOR THE YEAR 1767.

Publiſhed by ORDER of the

COMMISSIONERS OF LONGITUDE.

—————————————

LONDON:

Printed by W. RICHARDSON and S. CLARK,
PRINTERS;
AND SOLD BY
J. NOURSE, in the Strand, and Meſſ. MOUNT and PAGE,
on Tower-Hill,
Bookſellers to the ſaid COMMISSIONERS,
M DCC LXVI.

FIGURE 2.4   Title page from the first *Nautical Almanac*. (Royal Museums Greenwich ROG 7146 © National Maritime Museum, Greenwich, London.)

method of finding longitude as well as providing Britain with a comprehensive, authoritative national almanac. In December 1767 the Board of Longitude, pleased with the content, layout and usefulness of the work, authorised Maskelyne to continue to publish the *Nautical Almanac* on an annual basis, instructing that work proceed sufficiently quickly to ensure that the *Nautical Almanac* was published several years ahead so that they could be taken to sea on voyages of three or more years duration.[16] Maskelyne excelled at the task, and over the 45 years during which Maskelyne edited the *Nautical Almanac* it obtained a high reputation for accuracy.[17] Maskelyne ensured the accuracy and reliability of the tables in the *Nautical Almanac* by developing a national network of computers working in a highly systematic way based on his experiences with the first two editions.

[48]	APRIL 1767.			
	Diftances of ☽'s Center from ☉, and from Stars weft of her			
Days	Stars Names.	12 Hours.	15 Hours.	18 Hours.	21 Hours.
		° ′ ″	° ′ ″	° ′ ″	° ′ ″
1		40. 59. 11	42. 34. 44	44. 9. 51	45. 44. 35
2		53. 32. 7	55. 4. 24	56. 36. 16	58. 7. 45
3		65. 39. 18	67. 8. 27	68. 37. 14	70. 5. 39
4	The Sun.	77. 22. 36	78. 48. 58	80. 15. 1	81. 40. 46
5		88. 45. 20	90. 9. 27	91. 33. 21	92. 57. 0
6		99. 52. 6	101. 14. 34	102. 36. 52	103. 59. 1
7		110. 47. 42	112. 9. 6	113. 30. 25	114. 51. 40
6	Aldebaran	50. 36. 10	52. 4. 5	53. 31. 57	54. 59. 44
7		62. 17. 43	63. 45. 10	65. 12. 34	66. 39. 57
8	Pollux.	31. 25. 48	32. 53. 11	34. 20. 40	35. 48. 12
9		43. 7. 5	44. 35. 4	46. 3. 8	47. 31. 15
10		17. 51. 57	19. 20. 36	20. 49. 26	22. 18. 27
11		29. 45. 36	31. 15. 26	32. 45. 26	34. 15. 35
12	Regulus.	41. 48. 49	43. 19. 55	44. 51. 10	46. 22. 36
13		54. 2. 11	55. 34. 36	57. 7. 12	58. 39. 59
14		66. 26. 28	68. 0. 18	69. 34. 20	71. 8. 33
15		25. 4. 34	26. 39. 23	28. 14. 26	29. 49. 44
16	Spica ♍	37. 49. 37	39. 26. 14	41. 3. 5	42. 40. 8
17		50. 48. 40	52. 26. 59	54. 5. 31	55. 44. 15
18		64. 1. 2	65. 41. 3	67. 21. 18	69. 1. 48
19		31. 37. 14	33. 19. 7	35. 1. 13	36. 43. 32
20	Antares.	45. 18. 29	47. 2. 10	48. 46. 5	50. 30. 12
21		59. 14. 6	60. 59. 31	62. 45. 11	64. 31. 2
22		73. 23. 37	75. 10. 43	76. 58. 2	78. 45. 31
23	♑ Capri-	33. 17. 26	35. 4. 38	36. 52. 4	38. 39. 45
24	corni.	47. 41. 9	49. 29. 53	51. 18. 44	53. 7. 40
25	ɪ Aquilæ.	65. 57. 35	67. 29. 54	69. 2. 36	70. 35. 39
26		78. 24. 51	79. 59. 0	81. 33. 29	83. 7 45

FIGURE 2.5 Sample page from 1767 *Nautical Almanac*. (Royal Museums Greenwich ROG 7147 © National Maritime Museum, Greenwich, London.)

2.6 Computing the *Nautical Almanac*

It was essential that the tables in the *Nautical Almanac* were as error free as possible – a small error in one figure in a table could easily lead a ship's master to conclude he was 60 or more miles away from his true position with potentially fatal consequences.

Most of the tables contained in the *Nautical Almanac* could be compiled using a set of precepts (or instructions) for use with tables describing the calculation of the motion of a celestial body along with several other astronomical tables and a set of seven-figure logarithm and trigonometric tables.

Maskelyne developed a set of 'computing rules' – or algorithms as we would describe them today – for each of the tables published in the *Nautical Almanac* which he insisted his computers adhere to in their computations.[18] These rules were based on the methods Maskelyne had developed for the *British Mariner's Guide*, and most of them involved nothing more mathematically challenging than tables look-ups and seven or eight figure sexagesimal arithmetic to complete. Unfortunately only a few of the *Nautical Almanac* computing rules appear to have survived in the Royal Greenwich Observatory archives[19] – they were manuscript documents written as instructions to individuals when they started computing for the *Nautical Almanac*. Evidence from letters written by Maskelyne to his computers indicates that Maskelyne occasionally sent out updated rules from time to time as astronomical theory progressed, for example in response to developments in astronomical theory by Laplace, Lagrange, Burkahardt and Delambre, and others.[20]

To accompany the 'rules' Maskelyne sent his computers all the tables they need to complete the calculations. This list is daunting. While we do not know what books Maskelyne originally used, we do know that by 1799 new computers received 13 separate tables and a set of past editions of the *Nautical Almanacs*. The list, given in Table 2.1, illustrates that while the individual calculations may not have required complex techniques, computing the whole was a non-trivial matter.

The biggest practical problem in computing the *Nautical Almanac* tables lay in the large number of terms that had to be computed and the tedious nature of doing this

TABLE 2.1

List of Tables Sent by Maskelyne to Computers to Assist in Computing the *Nautical Almanac* (1799)

Tables from third edition of De Lalande's *Astronomy* 1792

Mason's *Lunar Tables*, 1780 (a replacement for Mayer's original tables used in previous *Nautical Almanac* editions)

Halley's 1759 *Tables of planetary motion*

Mayer's 1770 *Solar tables*

Michael Taylor's 1792 *Logarithms*

Michael Taylor's 1780 *Sexagesimal tables*

Hutton's 1781 *Tables of products and powers*

Hutton's 1794 *Logarithms*

Jean Bernoulli's 1779 *Sexcentenary tables*

Maskelyne's 1781 *Tables Requisite for use with the Nautical Almanac*

Set of previous editions of the *Nautical Almanac*

Three manuscript tables Maskelyne, or his assistants, had computed labelled V, XXXI and XXXII

Maskelyne's manuscript tables of equations of the moon, equations of equinoxes and obliquity of ecliptic

Manuscript tables of moon's distance from the sun and particular stars

---

Source: Diary of *Nautical Almanac* Work, Cambridge University Library, Manuscripts, RGO 4/324, folio 42.

repeatedly in sexagesimal arithmetic. For example, the calculation of the sun's longitude required 12 table look-ups – six of which required the computer to interpolate between existing tabular entries. There were then 14 seven- or eight-digit decimal or sexagesimal additions or subtractions to execute. Each calculation had to be repeated for each day of each month. The repetitive nature of the task coupled with the likelihood of wrongly looking up or copying down figures from other tables to use during the calculation meant that the risk of occasional errors being made was relatively high.

The system Maskelyne set up to assure himself that the calculations were correctly computed was essentially very simple. Each table in the *Nautical Almanac* was to be computed by two computers using the same algorithms and the same tables. One computer was referred to as the computer and the other as the anti-computer. Their work would then be compared and checked by a third person called, appropriately enough, the 'Comparer'. The comparer would examine the tables and search for any discrepancies between them. When a discrepancy was found, the comparer would make the necessary calculations to decide which computer had made the error and send it back to be recomputed.

The exceptions to this method were the tables giving the celestial longitude and latitude of the moon. For these the computer calculated the moon's position at noon, and the anti-computer computed the moon's position at midnight. When the comparer received the tables, he merged the two together and took fourth differences to establish any errors. A fourth difference greater than + or − 6 would indicate to the comparer that an error had been made in the last digit or greater in the computation. The comparer would then trace the error and write to the offending computer pointing out the error.[21]

In 1799 Maskelyne wrote out the tasks he expected his computers to perform.[22] While this document was written over 30 years after Maskelyne first set up the system, the procedure is unlikely to have changed significantly since the late 1760s.

1. Receive notification from Maskelyne which months to compute. Maskelyne usually allocated two months at a time and specified computing the moon's place either for noon or for midnight.
2. Calculate tables of the sun's position and motion, the eclipses of the satellites of Jupiter and the positions of the planets for each day of the month at apparent noon. (These corresponded to pages 2, 3, and 4 of a *Nautical Almanac* month.)
3. Calculate the moon's place at noon or midnight (*Nautical Almanac* month pages 5–7), whichever had been assigned.
4. Examine the computations by differencing, and correct any errors found.
5. Send the computations to the comparer.
    There then was a delay until the comparer had looked over the computations and located any discrepancies between the two sets.
6. Make any corrections required by the comparer.
7. Receive from the comparer the final table showing the moon's place at noon and midnight for each day of the month. On the same table was a list of the fixed stars to be used that month to calculate the required lunar distances.
8. Compute the latitude and longitude of the stars to be used for lunar distances for the days in question. Check these with the comparer.
    Here there was another delay while the calculations were checked.

9. When the stars' positions have been checked, calculate the lunar distance calculations for noon and midnight. (Some computers were then required to check these figures with the comparer before moving onto the next stage, but I have been unable to ascertain if this was so in all cases.)

10. Interpolate between the noon and midnight lunar distances to find lunar distances at three hourly intervals. (*Nautical Almanac* pages 8–11)

11. Calculate the exact times of a variety of astronomical phenomena, such as solar, lunar and planetary eclipses, from a list supplied by the comparer. (This data appeared on page 1 of each month of the *Nautical Almanac*.)

The duties of the comparer were similarly prescribed.

1. Receive notification from Maskelyne of who had been assigned to which *Nautical Almanac* month.

2. When received, compare the computations of the computer and the anti-computer. (Different computers worked at different rates, and there could be a delay of a couple of months or more between the reception of the two sets of computations for the same month.)

3. If a discrepancy occurs, calculate the entry to decide which computer needs to be corrected. Write to the computers pointing out their errors.

4. For the moon's latitude and longitude computations merge the two sets of data for noon and midnight together and take fourth differences to establish any errors.

5. Send corrected moon's places to both computers along with the names and dates of the fixed stars to be used to calculate lunar distances for that month.

6. Notify the computers of the likely astronomical phenomena (solar, lunar, planetary eclipses, etc.) for the month.

7. Prepare the copy for press using the more suitable computer's copy. (The computer's work which was neater and contained the fewer corrections was sent to the printer.)

8. Prepare the three pages of the *Nautical Almanac* which appear after the Preface containing the explanation of the symbols used, the signs of the zodiac, a list of feast days, an Oxford and Cambridge term list, and a list of solar and lunar eclipses for the year. Also prepare the list of astronomical phenomena for the first page of each month.

9. Compute the last page of each month (page 12), the configuration of the satellites of Jupiter.

10. Examine and correct the printed proofs.

This system of comparing and differencing calculations was set up to ensure that very few errors made their way into the published *Nautical Almanac*. To work in this way it was essential that the two computers assigned to a particular month worked independently. Indeed the system worked so well that it was able to detect when computers cheated. In 1770 the comparer, Malachy Hitchins, caught Joseph Keech (a clerk in the Lord Mayor's Court in London) and Reuben Robbins (a London schoolmaster) cheating. Keech and Robbins had copied each other's calculations and were discovered because Hitchins found that the two copies submitted matched too exactly. Maskelyne promptly sacked them and required that they paid for the extra work they had caused the comparer.[23]

2.7 A Network of Computers

Although Maskelyne managed the computing of the *Nautical Almanac* from the Royal Observatory in Greenwich and in the public perception the focus of work was Greenwich, the *Nautical Almanac* computers did not physically work at Greenwich. While no more than two computers worked on any one month of the *Nautical Almanac*, there were often 5 or 6 (and sometimes as many as 9) working on any one year and it would have been impossible to house them all in the restricted space at the Royal Observatory. Instead all the computing work was done in the computers' own homes and was often fitted in around other employment, with the *Nautical Almanac* work supplementing the computer's main income. The computers were paid on a piecework basis entirely based on the amount of work completed. One of the more prolific computers, Mary Edwards, was earning about £80 per annum in the 1790s, but unlike most of the other computers this was her only source of income and she worked full-time on the *Nautical Almanac*. For the first *Nautical Almanac* published for 1767 the rate of pay was £5 16s 8d per *Nautical Almanac* month computed, a rate which rose to £18 15s 0d in 1815.[24]

Different computers took different lengths of time to complete their work – some managed to complete a *Nautical Almanac* month within a calendar month, but others took perhaps 6 months to do each work month allocated to them. A rough average would be 6–10 weeks to complete the calculations, but there were delays while the comparer checked parts of the work and if the computer and anti-computer were not keeping pace with each other.[25] Maskelyne became very adroit at allocating *Nautical Almanac* months for computation so as to balance out the different lengths of time different computers took to return their work. For example, Mary Edwards, who was the fastest computer, would usually be allocated 12 months computing for each *Nautical Almanac* year and the corresponding anti-computer months shared out between four other computers.

Because the work was carried out in the homes of the computers, there was no need for them to be geographically close to Maskelyne or to the comparer who actually handled much of the day-to-day supervision. In light of the Keech and Robbins affair Maskelyne preferred to have computers working on the same month physically distant to each other. Just under half of the computers lived in different areas of London with the rest randomly spread out throughout England. At the beginning of the nineteenth century there was a slight concentration of computers in Cornwall where the longest serving comparer, Malachy Hitchins, lived and from where he recruited several local computers. Practically all communication between Maskelyne, the comparer and the computers took place by post. Several collections of such letters survive and illustrate that by the late eighteenth century a letter posted in Greenwich would often arrive in Cornwall the next day.

Most of the computers were recommended to Maskelyne by friends or colleagues. Sometimes members of the Board of Longitude made suggestions, and on other occasions individuals put themselves forward as possible candidates. What Maskelyne was looking for were people with some mathematical skill, patience, and attention to detail. Knowledge of astronomy could be learnt, and there is evidence on one occasion of Maskelyne recommending specific books to one of his computers to increase his understanding.[26] Much more important than an understanding of computational astronomy was the ability to look up tables accurately and to repeatedly perform arithmetic in seven or eight decimal or sexagesimal figures.

2.8 The Computers

The people that made up Maskelyne's eighteenth-century network of *Nautical Almanac* computers were a very varied collection of people ranging from vicars to housewives. The two things they all had in common were a need to supplement their income and an interest in astronomy and/or mathematics (Table 2.2).

A few of the computers, such as Charles Hutton and William Wales (Figure 2.6), were well-known mathematical practitioners of the eighteenth century,[27] but many others remain relatively obscure figures about whom very little is known. Many of the computers were self-taught in mathematics – only Malachy Hitchins and John Edwards had the benefit of a University education.

Thirteen of the computers were school masters. Some, for example William Wales, Charles Hutton and George Witchell, went on to hold prestigious teaching posts (at Christ's Hospital School, the Royal Military Academy at Woolwich and the Royal Academy at Portsmouth respectively), but most of the rest taught in small urban academies or in rural schools. Henry Andrews, one of the longest serving computers, ran his own school in Royston, Hertfordshire, where he was also a bookseller and purveyor of scientific instruments.

Four of the computers were clergymen. The Revd. Malachy Hitchins took holy orders in 1769 and held parishes in Cornwall alongside acting as *Nautical Almanac* comparer for 40 years. John Edwards too was a clergyman who was appointed as Preacher in Ludlow in 1778 where he and his family lived. Edwards' real passion was neither computing nor theology but making telescopes, and in 1784 he unfortunately poisoned himself with arsenic while he was refining the metals for making mirrors for reflecting telescopes. It was at this point that his wife, Mary Edwards, officially took over his computing work to support herself and her family, although it later transpired that actually she had been doing most of her husband's computing work

TABLE 2.2

Nautical Almanac Computers 1765–1811[38]

Henry Andrews	John Hellins	Thomas Sanderson
Charles Barton	Charles Hutton	Francis Simmonds
William Bayly	Nicholas James	Walter Steel
Thomas Brown	Joseph Keech	James Stephens
Richard Coffin	David Kinnebrook	Michael Taylor
John Crosley	Israel Lyons	Philip Turner
William Dunkin	John Mapson	John Wales
Richard Dunthorne	Richard Martyn	William Wales

FIGURE 2.6 William Wales (1734–1798), *Nautical Almanac* computer, astronomer on Cook's second voyage 1772–1775, Secretary to the Board of Longitude 1795–1798 and Master at Royal Mathematical School at Christ's Hospital 1775–1798. (Royal Museums Greenwich ROG B2625 © National Maritime Museum, Greenwich, London.)

all along.[28] Mary Edwards was the only female *Nautical Almanac* computer of this period until she introduced her daughter Eliza to the work in 1809.

Other computers were ex-Royal Observatory assistants trained by Maskelyne himself, and others were Board of Longitude astronomers who travelled with Cook and Vancouver on the Government-sponsored voyages of discovery which took place in the latter half of the eighteenth century. John Crosley had been Maskelyne's assistant at Greenwich before becoming a Board of Longitude astronomer. Between voyages he found stopgap employment computing for the *Nautical Almanac*. Other computers had employment as clerks or as surveyors.

2.9 Supporting the *Nautical Almanac*
Work through Observation

In addition to setting up a robust system for computing the *Nautical Almanac*, during his first year as Astronomer Royal Maskelyne set the pattern for an observing programme at the Greenwich Observatory. The observing programme was designed to both fulfil the terms of Maskelyne's Royal warrant to increase the accuracy of tables of the sun, moon, planets and fixed stars in order to assist in finding longitude at sea, and support and develop the astronomical tables on which the computations of the *Nautical Almanac* were based. The main priority for Maskelyne was ensuring that the moon was observed on every possible occasion that it crossed the Greenwich meridian.[29] Similarly the sun and planets were observed but with less priority. Data on star positions had already been collected by Maskelyne's predecessors and others, so Maskelyne restricted star observations to those near the celestial equator which could be observed in daylight and could be used to ascertain the accuracy of clocks. In addition there were a range of other observations such as solar and lunar eclipses, eclipses of Jupiter's satellites, comets, new planets and other special observations.

During his time as Astronomer Royal, Maskelyne had only one assistant employed by the state to live in the observatory to assist in both making the observations and reducing the raw observations for publication.[30] The post of assistant was poorly paid and rather lonely; assistants were required to remain single and not encouraged to socialise outside the Observatory. As a consequence over the years Maskelyne struggled to keep assistants; many considered their time at Greenwich as a suitable apprenticeship for more exciting or prestigious posts as astronomers on Board of Longitude-sponsored voyages of exploration. There is no question that Maskelyne trained many men who contributed to expeditions or projects sponsored by the Royal Society or Board of Longitude.

Maskelyne was concerned that the data gained from the observations made by his predecessors at the Royal Observatory had not been be published and in many cases had been removed from the Observatory, and was determined that his observations would be placed in the public domain with all speed. Through the Royal Society Maskelyne published his observations regularly.[31]

Maskelyne also collaborated closely with his opposite number in France, Joseph-Jérôme Lefrançais de Lalande (1732–1807) editor of the French almanac, *Connaissance des Temps*. From 1763 when they first met in England until Lalande's death in 1807,[32] Maskelyne and Lalande regularly corresponded and collaborated despite England and France being at war for much of that time. Indeed Maskelyne had used the ideas of Nicolas-Louis de Lacaille on precomputed lunar distances which Lalande had published in the 1761 edition of *Connaissance des Temps* as the basis for his design of the *Nautical Almanac*.[33] Maskelyne, or his assistants, compared the tables in the *Connaissance des Temps* against those prepared for the *Nautical Almanac* to check for discrepancies and errors in either or both of the publications. While the *Connaissance des Temps* did not then include the lunar distance tables, the comparison for the rest of the tables provided yet another assurance on their accuracy. But the collaboration between Maskelyne and Lalande went further. From the 1774 edition onwards the *Connaissance des Temps* included lunar distance tables from the *Nautical Almanac*. Although the explanation to the tables was translated into French, the tables themselves were identical to those published in the *Nautical Almanac* and therefore based on the Greenwich meridian. All the other tables in *Connaissance des Temps* took Paris as their base meridian.

The *Nautical Almanac* was highly effective in making the lunar distance method a practical way of finding longitude at sea. While initially intended for British navigators, by the mid-nineteenth century the *Nautical Almanac* was being used worldwide to find longitude – by either the lunar distance method or the chronometer method (for which you still required some astronomical tables supplied by the *Nautical Almanac*). The expeditions of Cook, Vancouver and others relied on the tables, and although Royal Navy navigators were initially slow to take up the new method, it soon caught on more widely in both naval and commercial shipping. Because Maskelyne managed to publish editions approximately five years ahead, it was possible to sell the *Nautical Almanac* not only throughout Britain but also in ports around the world.

The *Nautical Almanac* was also widely officially and unofficially reprinted. For example, by the early nineteenth century American reprints of the British *Nautical Almanac* were being printed in Massachusetts and being sold widely in US ports such as Boston, New York and Philadelphia.[34] But the *Nautical Almanac's* influence was wider that just the tables it contained. When the United States began to compute and publish its own almanac in 1852, the *American Ephemeris and Nautical Almanac*, the most controversial issue was which meridian to use. American astronomers and nationalists wanted the US almanac to use Washington as the meridian, but there was huge pressure from American navigators to use Greenwich as the meridian on which to base the *American Nautical Almanac*, as by the mid-nineteenth century, Greenwich was almost universally used as the international standard.[35] In the end a compromise was reached in which the astronomical parts of the *American Ephemeris and Nautical Almanac* adopted Washington as the prime meridian while Greenwich was used for the navigational section.

By the late nineteenth century Greenwich had become the de facto international prime meridian through the widespread influence and use of the *Nautical Almanac* which by now had annual sales of 20,000 copies.[36] The first International Geographical Congress in 1871 resolved that Greenwich should officially become the international meridian for longitude zero. There followed much international debate over the following ten years on the merits of having an internationally agreed prime meridian and the linked issue of having a prime meridian for an internationally recognised uniform standard of time in a world where continents were being increasingly linked by telegraph cables and crossed by railways running to timetables. These debates culminated at the International Meridian Conference held in Washington D.C. in 1884[37] which eventually agreed that Greenwich be officially adopted as the international prime meridian.

2.10 Conclusion

Maskelyne was intimately involved with both the chronometer and lunar distance method of finding longitude at sea. He played a crucial role in testing John Harrison's chronometers to see if they fulfilled the terms of the Longitude Prize and, by virtue of his appointment as Astronomer Royal, was a member of the Board of Longitude making decisions about Harrison's qualification for the prize. Maskelyne also built on improved tables of the motion of the moon developed by Tobias Mayer to turn the lunar distance method into a practical and cost-effective method of finding longitude at sea. While it was eventually decided that Harrison should be awarded the full prize, marine chronometers were expensive and difficult to make and very few ship's navigators could afford one at a cost of over £100 each. For many years only high profile explorers such as James Cook or ships operated by the East India Company could afford them. Even the Royal Navy did not carry them routinely, as navigational instruments were the private property of the ship's master rather than the Navy. Thanks to the work of watchmakers such as John Arnold and Thomas Earnshaw, who built on Harrison's work, chronometers gradually began to come down in price, but they were not universally affordable until the mid-nineteenth century. It was in providing a method of finding longitude at sea accessible to all that Maskelyne made his greatest contribution to navigation by the introduction of the *Nautical Almanac* – an annual publication that provided the predicted lunar distances needed to find longitude at sea. At a cost of 2s 6d per annum the *Nautical Almanac* was within the means of most navigators and hence was a practical solution to the Longitude Problem, and opened the door for Greenwich to be recognised as the international prime meridian.

3

George Biddell Airy, Greenwich and the Utility of Calculating Engines

Doron D. Swade

CONTENTS

3.1 Introduction

History has perhaps been unkind to George Biddell Airy (1801–1892). His failure to discover the planet Neptune was seen by his detractors as an egregious and unforgivable lapse and one that blighted a long and influential career. A new planet was of overwhelming public and scientific interest. Astrology, which was popularly held to be the key to fate and fortune, relied on astronomy with which it was entwined, to plot the motions of celestial bodies thought to choreograph human affairs. Observatories of the leading nations vied with each other for the coveted prize of new astronomical findings and to fumble the discovery of a new planet was to leave 'the gods in the sky'[1] unheeded. The loss of prestige, scientific and divinatory, was a matter of national shame. And Airy has been history's fall guy.

The circumstances of the Neptune affair are well documented. John Couch Adams, a 26-year-old Cambridge graduate, concluded, after laborious analysis, that irregularities in the orbit of Uranus might indicate the existence of a new planet.[2] In late September 1845 Adams, a junior scholar, modest and mild to the point of bashfulness, arrived without appointment to present his speculations to the formidable Airy, Astronomer Royal and master of the Greenwich Observatory, a position he occupied from 1835 to 1881.[3] Airy was away in France at the time. Two subsequent visits by Adams about a month later, also unannounced, were similarly frustrated. Airy was at dinner on the second visit, and the hapless visitor was refused entry, rebuffed by Airy's butler. An unproductive exchange of letters followed. Outside events overtook the stuttering exchange. During a trip to

Germany, a year after Adams' first approach, Airy learnt that the astronomer Johann Galle in Berlin, acting on detailed calculations made independently by the French astronomer-mathematician, Urbain Le Verrier, had found the planet on 23 September as Le Verrier had predicted. Airy's failure to look for the hypothetical planet on Adams' prompting was seen at home as a shameful failure of imagination. Adams' predictive feat has been described as 'probably the most daring mathematical enterprise of the century'.[4] Neptune was lost to the French, and Airy was vilified for what was seen as a grievous lapse and for squandering a unique opportunity for British science.

There were, it must be said, mitigating circumstances, professional and personal. Airy was exhausted by the burdens of office, especially by the demands of service on the Railway Gauge Commission. At the time of Adams' second visit in late October 1845, Airy's wife, Richarda, to whom he was devoted, was close to term at the age of 42 with her ninth child. Infant mortality was high, as were the risks of childbirth to a mother past the first bloom of youth. Airy was surely distracted with concern. To make matters worse a scandal erupted in the Observatory: a charge of incest was brought against an assistant there, and Airy had to deal with this distasteful business. The case culminated months later in the assistant's acquittal at the Old Bailey for the wilful murder of his incestuously conceived child. The affair broke at the time Adams ineptly sought audience with Airy, and threatened to damage the Observatory's reputation, which Airy's vigorous reforms and stern leadership had done much to restore. In the circumstances, inattention to an unassertive young Cambridge hopeful was perhaps understandable. Airy made no excuses despite the outcry. By his reckoning searching the skies for hypothetical planets was outside the mandate of the Observatory, a government-funded body established specifically to aid astronomy and navigation through observation. Theoretical astronomy, which included the speculative existence of new planets, was outside his narrow interpretation of his professional responsibility. The construal of his role and that of the Observatory as strictly utilitarian was not conjured for the occasion to evade recrimination. He had clearly articulated the Observatory's public obligations as essentially practical, in the service of which he was accountable to the Board of Visitors responsible for overseeing his stewardship. A heavenly wild goose chase, whatever the prize, might invite censure.

In the words of his son Wilfrid, Airy 'had the greatest dread of disorder ... even in the smallest matters'.[5] It is clear from his metronomic habits, his meticulous document-filing and information retrieval systems, and his preoccupation with procedural discipline that Airy was obsessed with efficiency, organisational detail, and record-keeping. The need for order and method is portrayed as Airy's dominant character trait, and in his pragmatism we find an exquisite instrument of prevailing scientific values. His emphasis on practical observation and the grindingly systematic reduction of observational data to produce tables for astronomical navigation was part of a broader movement in which scientific speculation was regarded as irresponsible, and something to be crushed by 'astronomical book-keeping and precision'.[6] In his embrace of a narrowly functional role we find a convergence of temperament and contemporary scientific culture.

At the time of the Neptune affair Airy was sufficiently well established to defend his conduct to his masters. He weathered the storm of blame, criticism, and bitterness that followed in the immediate aftermath. History has been less forgiving. The accusations never entirely faded and Airy has been cast, perhaps irredeemably, as a prototype of the uninspired bureaucrat whose love of procedure and whose failure of imagination cost British science dear.

3.2 Errors and Misconceptions

The Neptune episode is prominent in the politics of nineteenth-century science. There is another less publicised affair, in which invention and new ideas suffered a similar fate at least partly, and possibly largely, as a result of Airy's seemingly pedestrian outlook and the high value he placed on the mundane.

Charles Babbage (1791–1871), mathematician and polymath, designed vast automatic calculating engines that represent the most ambitious attempts to mechanise computation in the nineteenth century. Babbage failed to complete any of his engine designs in physical form, despite decades of design and development, advanced engineering, private wealth, liberal government funding, and the social advantages of a gentleman of science with access to the highest echelons of society. Airy was consulted by government on at least three occasions for his views on the utility of automatic calculating engines. In 1842, a critical time for the survival of the Babbage's engine project, when invited by government to give an opinion on the engine, Airy pronounced it 'worthless'.[7] As Astronomer Royal, the highest office in civil science, and *de facto* scientific adviser to government, Airy was the official arbiter of utility, and his views had a defining influence on the fate of Babbage's engines.[8] Airy's single damning judgement is frequently quoted to reinforce his image as a stereotype of bureaucratic dullness, a mediocre Salieri who used his position to thwart genius of which he was himself incapable (Figure 3.1 and 3.2).[9]

FIGURE 3.1   George Biddell Airy (1810–1892). (Illustrated London News, 4 January 1868.)

FIGURE 3.2 Charles Babbage (1791–1871). (National Portrait Gallery P28, circa 1847-51, © National Portrait Gallery, London.)

Babbage, whose incontinently savage public attacks were the despair of his friends, published accusations that he was a victim of a conspiracy of which Airy was part and that the Astronomer Royal's hostile counsel to government was biased by malice.[10] Yet Airy's one-word dismissal was not an isolated irritable aberration. With one notable exception Airy consistently rejected the practical and economic utility of automatic calculating engines and several other initiatives for mechanical calculating aids. One casualty of Airy's persistent dismissals was a promising set of mathematical ideas, and we perhaps owe it to history to excavate Airy's views, examine his argued justifications, and defer, at least for the meanwhile, joining the chorus exalting Babbage at Airy's expense.

Printed mathematical tables were indispensable to scientists, astronomers, navigators, engineers, bankers, insurance companies, journeymen, and anyone who wished to do more than trivial calculations. However, tables, which were calculated, typeset, and printed manually, were thought to be riddled with errors, and the issue of errors, their seriousness and extent, is one that dominates historical accounts of automatic calculation in the nineteenth century. It is accepted that eliminating the risk of errors in tables' production was the principal purpose of Babbage's engines and those of other inventors of automatic calculating aids. The 'unerring certainty of mechanical agency' would ensure absolute accuracy. Machinery would come to the aid of fallible humans. Or so the engine advocates would have the world believe.[11]

It is Dionysius Lardner more than anyone else who is responsible for positioning errors at the centre of the debate. Lardner was a prolific populariser of science and gifted lecturer, a colourful figure in the early Victorian scientific landscape. In an energetic article of numbing length, published in 1834, Lardner takes pains to establish that printed tables were riddled with errors, irreparably so, and in enthusiastic promotion of Babbage's work he argued for Babbage's engines as the solution. Lardner's article is the most comprehensive contemporary account of Babbage's Difference Engine project and has had a defining influence on the historical perception that has prevailed since that time – that errors in printed tables were not only the jumping-off point, but the enduring motive for Babbage's efforts.

However, the chronology of events and the circumstances of the article invite a revision in the accepted perceptions of the role of errors. Lardner's article was published in July 1834, some 13 years after Babbage's first conception of his machine.[12] Whatever the effect of Lardner's article on historians since, it can have had no material influence on Babbage's original ideas, which pre-date Lardner's interest in the subject by over a decade. To uncouple the backward projection of later times onto the events of the day, it is instructive to review Babbage's earliest writings recorded close to his original conception of the engine and before justificatory rhetoric began to skew the arguments.

The papers that are most revealing of Babbage's earliest notions were written during the last six months of 1822 following trials using a small working model completed in the spring of that year. What is immediately noticeable from the suite of five papers is that the dominant preoccupation is not the production of error-free tables, but mathematics. While the elimination of errors does feature, it does so alongside a number of mathematical interests, and errors are not nearly as prominent in these early writings as post-Lardner accounts suggest. In some of the papers, errors are barely mentioned. When they are, it is mainly as a device of persuasion in attempts to justify, in an open letter to Humphrey Davy, President of the Royal Society, the benefits of the engines on the grounds of practical usefulness.[13] While the prospect of the engines offered remedies for known deficiencies in the production of printed tables (speed, generality, and correctness), what preoccupies Babbage in these papers are new implications of the engines.

Amongst the ideas Babbage explored is that of computation as systematic method e.g. for solving equations. He described how cycling the machine constituted a definite procedure for finding the roots of equations for which there was no known analytical solution, an idea that elevates computation from the banal chore of calculation to procedural method. The engine represented a new technology for mathematics, and this was virgin territory. He saw the potential of the engine to calculate the nth term of a series for which there was no generalised expression as well as the heuristic value of the machine to suggest new series. He also predicted the eventual dependence on machines for scientific calculation and foresaw new branches of mathematics to optimise the efficiency of machine computation. It is clear from these early writings that there were theoretical and logical dimensions to Babbage's aspirations for his engines that transcend the practicalities of reliable tabulation. These were new ideas, but off-piste when it came to the analytical conventions of contemporary mathematics.[14]

For the engine advocates the machines promised theoretical and practical benefits: a solution to the problem of correctness, supply, and speed in the production of printed mathematical tables, as well as the prospects for new areas of mathematics.[15] How, in the face of such promise, did Airy justify such vigorous opposition to the utility of the machines?

3.3 A Delicate Matter

Airy was a relative latecomer to the fray. His first involvement as adviser to government on the utility of Babbage's engine was in September 1842, some twenty years after the start of Babbage's efforts to construct his first full-sized engine and ten years after the collapse of his major construction project when, in 1833, his engineer, Joseph Clement, downed tools. The lead-up to Airy's first involvement was a carnival of confusion and frustration. For ten years Babbage had conducted a desultory exchange with government in an attempt to resolve the tangle of commitments between himself and the Treasury following the debacle of Clement's walkout. His letters are at times irritable, aggrieved, enraged, impatient, and indignant. The technical content is often unclear and apparently self-contradictory. In an oblique way he seemed to advocate abandoning the existing engine, half-completed at massive Treasury expense, in favour of a new more sophisticated machine that could be finished more cheaply but with vague, unspecified, and contradictory benefits. At the same time he seemed conscious of an obligation to complete the old. Small wonder that the Treasury was at a loss as to how to reply. It was far from clear what question it was Babbage was entreating them to answer. The political circumstances were volatile, and the years of negotiation were not helped by the game of musical chairs being played at Westminster with nine changes of government in the 20 years since the start of the project, six Tory and three Whig. The sclerotic delays, confusion, and frustration are thought to have provided the parodic model for Dickens' Circumlocution Office in *Little Dorrit*.[16]

The fraught and wearying process was brought to a head by a series of increasingly importunate letters from Babbage to the Prime Minister, Robert Peel, written between August and October 1842. Peel, under pressure, sought advice. 'What shall we do to get rid of Mr. Babbage and his calculating machine?' was his mock lament to William Buckland, a geologist to whom he turned for advice from time to time. 'What', he asked, 'do men really competent to judge say in *private*?'[17] He was candid with Buckland that he would prefer to cut his losses and be shot of the project which he saw as haemorrhaging money with no end in sight. Buckland's reply is not recorded, and Peel appears to have charged his Chancellor, Henry Goulburn, to find out what was being said behind closed doors.

Goulburn sought the views of John Herschel and enlisted Airy's help in securing Herschel's counsel. It is not clear why Goulburn saw fit to use Airy as an intermediary.[18] If Goulburn was lining up someone hostile to the engine venture, Herschel was an unlikely choice: he had been a close friend of Babbage since their undergraduate years together at Cambridge and had served on all three Royal Society Engine Committees convened to assess the utility, feasibility, and progress of the machines.[19] He also managed the engine project during Babbage's year-long absence when, in 1827, close to breakdown following the deaths of his wife, father, and a new-born son within the year, Babbage went on a recuperative trip to Europe.[20] In choosing someone close to Babbage, and someone both supportive and knowledgeable about the engines and their history, the government was fairer to Babbage than he was later willing to admit.

The approach to Herschel was made through Airy. Goulburn wrote to Airy on 15 September 1842, asking him to secure Herschel's views on the wisdom of expending any further public funds to complete the unfinished engine. Goulburn ended his letter with a casual comment, seemingly as an afterthought: 'If you could add your own [opinion] also it would be conferring on me an additional favor'.[21] Airy seized on the offhand invitation. He replied immediately, and his response reveals some of the tensions and resentments between the lines of the official minutes.

3.4 The Astronomer Royal's Verdict

Airy first discredits the favourable findings of the Royal Society. He alleges that the specially convened engine committees were populated with Babbage's acolytes whom he described as 'all private friends and admirers of Mr. Babbage ... a little blinded by the ingenuity of their friend's invention'.[22] It is the case that of the 12 members of the 1823 Committee at least seven were Babbage's supporters of one sort or another – close friends, men of science involved with Babbage in collaborative projects, or sharing his 'declinist' sympathies. Despite this home-side padding it seems that all were not uniformly enamoured of the scheme. Airy reported to Goulburn that the Committee's report, while 'very favourable', was not unanimous. 'It was boldly stated by Dr. Young ... that if finished, it [the machine] would be useless'.[23] Airy does not state the basis of Young's objection, but his description of Babbage's reaction is damning[24]:

> For this [the statement that the engine would be useless], he [Young] was regarded by Mr. Babbage with the most intense hatred ... Mr. Babbage made the approval of the machine a personal question. In consequence of this, I, and I believe other persons, have carefully abstained for several years from alluding to it in his presence. I think it likely that he lives in a sort of dream as to its utility.

Airy then seeks to correct the false notion that the Difference Engine was intended for general-purpose calculation:

> An absurd notion has been spread abroad, that the machine was intended for *all* calculations of every kind. This is quite wrong. The machine is intended *solely* for calculations which can be made by addition and subtraction in a particular way. This excludes all ordinary calculation.[25]

He then demolishes the idea that the engine would be of any practical use at all:

> Scarcely a figure of the Nautical Almanac could be computed by it. Not a single figure of the Greenwich Observations or the great human Computations now going on could be computed by it. Indeed it was proposed only for the computation of new Tables (as Tables of Logarithms and the like), and even for these, the difficult part must be done by human computers. The necessity for such new tables does not occur, as I really believe, once in fifty years. I can therefore state without the least hesitation that I believe the machine to be useless, and that the sooner it is abandoned, the better it will be for all parties.

Airy's rejection of the utility of the machine is robust and complete. He made no concessions to the finer feelings of those who supported the machine. The brusque and bruising tone suggests strong feeling – resentment, indignation, and even anger – and his *ad hominem* portrayal of Babbage as irrationally defensive to the point of social dysfunction is hardly charitable.

The bluntness, though not the anger, in Airy's letter is a feature of his writing in general. For the Astronomer Royal it was grit rather than grace. This is true of his scientific writing and business letters, as well as personal correspondence. In a 'Personal Sketch' of his father, Wilfrid Airy observes that in all matters 'he kept his object clearly in view, and made straight for it, aiming far more at clearness and directness than at elegance', writing with 'great ease and rapidity'.[26] Of the countless letters from Airy to his wife, Wilfrid comments that 'they are not brilliantly written, for it was not in his nature to write for effect … but they are straightforward, clear and concise'.[27] Nor did Airy shrink from confrontation. In the testimony of his son 'he never hesitated to attack theories and methods he thought scientifically wrong' and 'in debate and controversy he had great self-reliance, and was absolutely fearless'.[28] Airy was evidently a formidable opponent and perhaps not too much should be made of the robust combative style of his letter to Goulburn. Even so, the views he expressed are unmistakably strong, especially for someone so evidently disciplined, considered, and possessed of 'great steadiness of temper'.[29]

Goulburn's letter is marked 'Private and Confidential', and all subsequent exchanges are headed 'Private' by both parties. Such was the sensitivity of the matter that Herschel sent his report sealed to ensure 'perfect insulation of minions'.[30] It might be thought that Airy's pen was freed by the protection of strict confidentiality and that the excesses of his attack were licensed by insurance against scrutiny by others. Not a bit of it. He seemed to be goading Goulburn into wider disclosure, writing that Goulburn was 'fully entitled, in courtesy as well as in right, to use my expressed opinion in any way that you shall think fit'.[31] Goulburn was evidently gratified by Airy's response, whose condemnation fortified Peel's view that the engine was a 'very costly toy' and also endorsed Goulburn's reluctance to proceed.[32] He replied by return to Airy tendering his 'best thanks for your satisfactory letter'.[33]

There are a number of oddities in Airy's account of the 1823 Royal Society meeting. The description of Young's dissension gives the impression that an acrimonious exchange took place between Young and Babbage and that Airy witnessed it. However, it is almost certain that neither Airy nor Babbage was present. Neither is listed as a member of the Committee. Airy was a 21-year-old Cambridge undergraduate at the time, and there is no evidence that Babbage, as an interested party, participated in person in the Committee's business as a petitioner or sponsor.[34] Babbage's diary entries show that he was kept closely informed about the proceedings of the Committee, probably through hearsay reports from his supporters, prominent amongst whom were Baily, Gilbert, and Wollaston, but in his diary he made no reference to any altercation with Young.[35] He mentions Young's dissension but casts a very different light on the degree of cordiality, and also on the level of Young's support which, in the final report, was unexpectedly fulsome.[36]

The evidence suggests that no altercation took place between Babbage and Young at the meeting. Babbage's diary was written at the time of the events, Airy's nearly twenty years later, both from hearsay reports. It seems likely then that the hostility of Airy's outburst had its roots in events that occurred in the intervening years and projected backwards by Airy onto an episode twenty years earlier. There was much in Babbage's behaviour to have riled Airy in the meantime. There was an ill-tempered competition in 1826 for the Lucasian Chair of Mathematics at Cambridge during which Babbage allegedly threatened legal proceedings (Airy routed Babbage by eight votes to two); widely felt resentment of Babbage's neglect of his professorial duties when, in 1828, he did secure the Lucasian Chair (this in contrast to Airy's exemplary industry during his own incumbency); and Babbage's role as an 'expert witness' in favour of Sir James South against Airy and the Reverend Richard Sheepshanks in the rancorous legal dispute – "the astronomers' war" – over an allegedly faulty telescope mounting.[37] But at the time of the first Committee meeting on the Engine in 1823 there was little hint of tribulations to come.

Meanwhile Herschel was squirming over his brief. His sealed report, sent to Airy for onward transmission to Goulburn, is long and elaborate. In it he steered an uncomfortable path between loyalty to Babbage and an apparently genuine belief in the utility of the machine on the one hand and public duty on the other. The document was clearly well-meant, but circumspection weakened his stated conviction in the benefits of the machine. It seems that Goulburn's final decision was influenced more by Airy's short clear damnation than by Herschel's cautious support: Goulburn wrote to Babbage on 3 November 1842 axing the project on the grounds of cost.[38]

In his *Autobiography* Airy later recounted his dismissive judgement:

> On Sept. 15th Mr. Goulburn, Chancellor of the Exchequer, asked my opinion on the utility of Babbage's calculating machine, and the propriety of expending further sums of money on it. I replied, entering full into the matter, and giving my opinion that it was worthless. – I was elected an Honorary Member of the Institution of Civil Engineers, London.[39]

This is the most frequently quoted of Airy's condemnations. It despatches half a lifetime's work in a brusque dismissal without supportive argument or self-justification. The throwaway mention of his election to the Institution of Civil Engineers appears to add insult to injury, ranking, as it seemingly does, an honorary membership of a professional institution with Babbage's grand venture, the conjunction of the two statements a sign of disregard. We should perhaps not read too much into the apparent brutality of the entry. The *Autobiography* has the format of an annual diary with sections for each year from 1836 to 1891, and abrupt transitions between unrelated events are not uncommon. The entry for 1826, for example, reconstructed by Airy from his personal papers, memorabilia, and the quires that served as a form of scratch-pad for his daily workings, juxtaposes a record of the tutor's stipend (£50) and the ellipticity of heterogeneous spheroids.[40] Starker *non sequiturs* from the same year conjoin Airy's commencement of tuition in Italian, his witnessing a murderer guillotined in the Place Martroi, and investigations into pendulums and the calculus of variations. The sources on which the *Autobiography* is based are largely, but not exclusively by Airy's hand, edited and supplemented by his son, Wilfrid, after his father's death.[41] Perfunctoriness is to some extent a feature of the genre, and in the case of any given example of disjointedness it is impossible to know whether this is a deliberate jump-cut by Airy himself or an artefact of Wilfrid's editing.[42]

Airy's opposition to the engines was known much earlier. Thomas Romney Robinson, an Irish astronomer, wrote to Babbage in 1835, the year Airy took up his appointment as Astronomer Royal, to tip him off about a conversation to which he was privy in which Airy, in company, had pronounced the engine 'useless' on the grounds 'that if the money spent on it had been applied to pay computers, we could have had all that is wanting in the way of tables'.[43] In a piece of contemporary report-age the actor William Macready, whose diaries provide an inexhaustible source of society gossip, commented on Babbage's engine: 'Professor Airy says the thing is a humbug. Other scientific men say directly the contrary'.[44] The Astronomer Royal's views on the Engine were clearly all about town.

3.5 The Case against the Engines

If we disinfect Airy's views of the engine from his personal feelings towards Babbage, we can identify a set of technical assertions, supposing for the while that the technical objections were not invoked simply to support a prejudice. The first is that the machine was not a general-purpose calculating device for 'ordinary calculation' but intended for the specialised purpose of tabulation (using repeated addition and subtraction as required by the method of differences). Airy is here correcting a misapprehension in public perception rather than contradicting any claim made by the engine advocates. His second objection is that the machine was irrelevant to the ongoing computations at the Greenwich Observatory which he superintended, and for the Nautical Almanac. The production of such tables was the Observatory's daily staple, and Airy's views, right or wrong, carried a great deal of weight. The third objection is a compound one: the extreme rarity of the need for new tables of standard functions such as logarithms, and that 'even for these, the difficult part must be done by human computers'.[45] Given that the engine enthusiasts had based their advocacy on the growing demand for new tables and on the guarantee of accuracy promised by machinery, Airy's point is dam-aging. The starting-point of any machine calculation relies on manually computed starting values as well as the accurate use of formulae. Both the calculation of starting values and the calculation of the maximum number of results for which the approxi-mation remained sufficiently accurate to ensure last digit accuracy were susceptible to human error. The reliance on manual calculations was a weakness in arguments for the infallibility of sub-tabulation by machine and one that the engine advocates did not go out of their way to advertise. But the point made by Airy does not refer directly to errors but to the 'difficulty' of the preparatory calculations. He is almost certainly referring here to the high level of abstract analysis and complex calculations needed to ensure that the results for a given run of machine calculations remain within the requisite limits of precision. Airy implies that the burden of such preparation out-weighs the benefits of sub-tabulation by machine.

In his report to Goulburn Airy does not elaborate on the specific objection raised by Young. However, the objection appears in the record: 'that it would be far more useful to invest the probable cost of constructing such a calculating machine ... and apply the dividends to paying calculators'.[46] Young's argument, shared by Airy (as reported to Babbage by Romney Robinson), was that there was no economic case for the machine.[47]

Airy's report to Goulburn provides clues about Airy's imaginative horizons. There is no reference to mathematics or computational theory of the kind that Babbage envisaged in his references to the engines. Yet he was well placed to judge whether these ideas had merit. Airy had excelled at mathematics at Cambridge, coming first in each of his undergraduate years. He recalled that when the second-year results were posted, his name was separated from the rest by two lines, indicating that he had scored double the marks of his nearest rival.[48] He graduated in 1823 with top honours – Senior Wrangler and first Smith's Prizeman. As he later recalled the occasion: 'rarely has the Senate House rung with such applause as then filled it. For many minutes, after I was brought up in front of the Vice-Chancellor, it was impossible to proceed with the ceremony on account of the uproar'.[49] Equipped as he was, it is curious that he failed to register, let alone acknowledge, the mathematical potential of the engine as articulated by Babbage in his published papers. It seems that his vision was limited to little beyond the practical capacity of the engines to compete with the routine clerical and computational practices at the Observatory and that the daily staple of tabulation – producing tables of the positions of the moon and stars based on daily telescopic observation – served as the bench-mark of utility.

In his emphasis on economics, Young, it seems was cut from a similar cloth. He appears less concerned with new applications than he is with an economic justification for replacing existing practices by machine. Young was not given to writing reports of fulsome excess even in favour of projects he supported. According to a contemporary view, he 'was not so easily seduced into enthusiasm'.[50]

3.6 A Personal Vendetta

In 1851, the year of the Great Exhibition, Babbage launched a public attack on Airy accusing him of being party to a personal vendetta against him and of influencing the government against the engines as part of a calculated scheme to discredit him. In his book *Exposition of 1851*, Babbage alludes to dark forces acting against him in collusion with government and names Airy as his villain. The allegations follow a tortuous logic: that acting for his friend, Sir James South, in a court case South brought against the instrument makers Troughton and Simms, Babbage had opposed the Reverend Richard Sheepshanks, a close friend of Airy's, and that Airy, loyal to Sheepshanks, biased his advice to government against the engines to avenge Babbage's offence. Babbage accuses Airy of bearing him a grudge and of allowing malice to corrupt his professional views.

Exposition is less a work of analysis than an angry assault in which Babbage's sense of injustice is infused with bitterness. The mix is one of unmistakable misery. He was advised by friends to avoid *ad hominem* attacks or not to publish. In the Introduction he defies them. The work has been described as 'the diatribe of a disappointed man ... disfigured by personal allusions, in giving utterance to which, he wronged his better nature'.[51]

The *dramatis personae* seem to belong more to a Punch-and-Judy show than to gentlemanly society. South has been described as 'irascible almost beyond the bounds of sanity when engaged in controversy' and had a reputation for being an unpleasant maverick.[52] Sheepshanks was a belligerent bruiser with legal training and always on the lookout for a scrap. He openly admitted that he held Babbage in 'great contempt', sought to intimidate him, and that he (Sheepshanks) was out to get him.[53] In 1839 the court found against South in favour of Troughton. South smashed the telescope and ostentatiously auctioned the pieces. Sheepshanks died in 1855. Not content to let matters rest, South dispensed with the conventions of posthumous eulogy and proceeded to pursue Sheepshanks beyond the grave by attacking the Astronomical Society obituary which, true to the genre, was predictably favourable to Sheepshanks.[54]

By 1851, the time of Babbage's accusations, Airy's professional standing was unassailable. He brushed off the attack, disdaining to make any public defence. For amusement he parodied the tortured progression of Babbage's allegations against him in a prose poem modelled on *The House that Jack Built*.[55] Airy has so far appeared as something of a dry stick, but the humour and irony of the verses show him in a more human light. He was entirely secure and could afford to make a mischievous joke at Babbage's expense. The seventh of eight verses reads:

> These are Treasury lords, slightly furnished with sense,
> Who the wealth of the nation unfairly dispense:
> They know but one man, in the Queen's vast dominion,
> Who in things scientific can give an opinion:
> And when Babbage for funds for the Engine applied,
> They called upon Airy, no doubt,* to decide:
> And doubtless adopted, in apathy slavish,
> The hostile suggestions of enmity knavish:
> The powers of official position abused,
> And flatly all further advances refused
>
> > For completing the Engine that Charles built.[56]
>
> Foot note
> *Upon this bare assumption, it seems, as a base,
> Mr. Babbage has founded the whole of his case;
> But I have not remarked, in his book, a citation,
> That gives to this notion the least confirmation.

Airy's self-portrayal is that of a public servant of scrupulous impartiality who, with one late exception, appears offhand in his interest in the calculating machines. The vast archive of his papers confirms their author as prolifically productive and conscientiously dutiful. Yet a letter from William Whewell to Herschel written in October 1822, shortly after Babbage's first published announcement of his invention, signals what may have been the seeds of later rivalry:

> You have of course heard from Peacock about Airy, a pupil of his, and certainly a man of very extraordinary talents. The reports about Babbage's machine have, it seems, excited him to attempt something of the same kind. He and another man have made a machine to solve cubic equations, but besides this he has, so far as I can make out, invented a good deal in the way of Babbage's contrivance. He is not here at present, but his friend tells me that it has got toothed wheels, working in one way for the differences and in another for the digits. If it be a similar invention to that, it is probably an independent one; for I do not know any way by which he has got any lights about Babbage's affair.[57]

Airy's scratch-pad journal shows a 'sketch for a computing machine (suggested by the publications relating to Babbage's), [and] sketch of a machine for solving equations'.[58] Though Airy maintained an inventive interest in instrumentation, locks, and other mechanical contrivances throughout his career, there is no evidence yet that he developed his early ideas on calculating machines or that he had any ambitions of his own to design or build one, either as an intellectual exercise or for practical purposes. On the face of it at least it seems that his personal interest in the engines started and stopped with a doodle on his student notepad while a Cambridge undergraduate.

Airy was not alone in his opposition to the engines. In 1844 Nils Seelander, a Swedish astronomer, argued that the standard mathematical functions (logarithms and trigonometric functions) had already been computed 'with such accuracy that they leave little to desire' and that effort would be better directed towards improving the accuracy of observational data on which astronomical and other physical tables depended.[59] Le Verrier later similarly rejected the completed and working Difference Engine by the Scheutzes as offering any worthwhile practical benefit to the work of the Paris Observatory.[60] Two distinguished Continental astronomers shared Airy's antagonism to the utility of the machines.

3.7 The Tale of Georg and Edvard Scheutz

In the decades that followed Airy was consulted by government on three further occasions for his views on automatic calculating machines. He was also petitioned by individuals seeking scientific endorsement for new mechanical calculating aids and devices. In the case of the three government consultations the objects of interest were the difference engines by Georg Scheutz (1785–1873), a Stockholm printer, publisher, and journalist, and Edvard (1821–1881), his son.[61] Airy's views, recorded in correspondence with government and with hopeful petitioners, give insights into the generic criteria by which he judged the utility of machine-assisted calculation. With one exception his pronouncements were consistently negative.

Ignited by Lardner's 1834 article on Babbage's engine the Scheutzes designed and built an automatic printing calculator based on the same mathematical principle proposed by Babbage, but with mechanisms of their own design. The wooden-framed prototype, built in Stockholm, was finished in 1843 and the Scheutzes, looking for buyers, offered a fully engineered version to the British government. Airy was a member of the advisory committee assembled to counsel the Home Secretary, Sir James Graham. The record of the committee's deliberations has not been found, but the outcome is known: the proposal was rejected on the grounds that Parliament would be unlikely to support a foreign invention of the same kind as the English one, which had already proved both problematic and extravagantly expensive. It was after all only a year since Peel on Goulburn's advice had written off the vast investment in Babbage's engine. They had already been badly burnt by Babbage.

The Scheutzes were persistent. In 1854 they brought a fully engineered version to London, where it became an object of learned curiosity. Airy was omitted from the four-man Royal Society committee which reported respectfully but unenthusiastically. But he contrived to view the machine privately and volunteered his views in the *Philosophical Magazine* which had published the Society's report.

In his letter to the editor Airy asserts that 'the demand for such machines has arisen on the side, not of computers, but of mechanists'.[62] The allegation is clear: that the stimulus for the machines derived not from the deficiencies in existing computational methods but from the enthusiasm or entrepreneurial motives of technologists. He also implies with polite disdain that the engine advocates were ignorant of the real needs of tabular calculation. He then lists four stages typical of the most common computations used in manual tabulation and examines the respects in which each might or might not benefit by mechanisation. Using lunar tables as an example he argues that manual calculation of the widely spaced pivotal values that serve as the starting values 'can never be dispensed with', i.e., that the starting points of the tabulations are irreducibly reliant on human computers and are not amenable to machine calculation. The second stage was verifying the pivotal values by differencing (checking by repeated subtraction of successive pairs of values). This he concedes is 'well adapted to mechanical action'. This suggestion is remarkable. Airy proposed that rather than using the machine for generating tables *ab initio* by repeated addition as described by the engine advocates, it could be used instead to verify already computed tables by repeated subtraction. Verification by differencing was a well-established technique in manual computation but nowhere did Babbage mention verification by mechanical differencing. Computer pioneer, Maurice Wilkes, is incredulous that Babbage could have missed this.[63] In the light of twentieth-century tabulation practice Airy's suggestion is original, ingenious, and farsighted.[64] Airy concedes that the third stage (the preparation of first and second differences) could benefit from machine computation, but the simplicity of the calculations barely warrants it; the fourth stage (sub-tabulation by differences) could similarly benefit, but he implies that the Scheutz difference engine was more elaborate than necessary for this purpose.

The views Airy reveals here were the first public expression of reasoned arguments for his opposition to the machines, this some 14 years after his private condemnation to Goulburn. This time he signed the letter using the form he reserved for documents written in his official rather than his private capacity.[65] He evidently wished to register his scepticism as an official position. With the exception of the new suggestion of verification by mechanical differencing, his assessment of the engine was soberly discouraging and an unambiguous signal that belief in its utility was ill-informed or otherwise misguided.

Airy's published views on the Scheutz engine were unsolicited and based on his private viewing of the machine. But there were two further instances in which his opinion on the Swedish engine were actively sought by government. The first was in 1857 when Airy was directly consulted by the Treasury to advise on the purchase of a Scheutz difference engine at a cost of £1,200 for use in the General Register Office (GRO).[66] The proposal for the purchase was made at the joint instigation of Edvard Scheutz and William Farr of the GRO, the latter a leading statistician. Farr saw the engine as a means to aid the production of an English Life Table expanded to include interest and annuities for rates other than the standard and stable 3%, life assurance rates and yields for several variations of joint lives, and a set of life tables by district to identify the relative effects of rural and urban environments on mortality. Airy was asked to assess the likely utility of the engine not just at the GRO for Farr's project on the 1864 Life Table, but also in two other government offices: Greenwich and the Nautical Almanac (Figure 3.3).

FIGURE 3.3 The Scheutz Difference Engine by Bryan Donkin for the General Register Office, 1859. (Science Museum 10303296, courtesy of Science and Society Picture Library.)

If the views Airy expressed to Goulburn blurred the personal and professional, his dealing with his new commission was ostentatiously procedural. Airy prepared a survey of ten questions and sent near-identical versions to John Russell Hind, Superintendent of the Nautical Almanac, and to Sir George Graham, Registrar General.[67] The set of questions used, as a datum, the existing manual practices in each office and represents the most searching survey yet of the operational utility of a calculating engine. He asks specifically for the frequency with which specific mathematical functions were generated using differencing techniques of the fourth order; the need for the automatic production of stereotype plates; whether the loss of intermediate results (a feature of the machine technique) was a disadvantage; the frequency with which temporary as distinct from permanent tables were required; whether if the machine were infrequently used, it would be preferable to employ a computer on secondment or someone specific to maintain the machine; and what mechanical aids (slide rules, adding machines and the like) were currently in use, and so on. To Hind at the Nautical Almanac he sent supplementary questions: an invitation to comment on Airy's published suggestion to use the machine for verification by repeated differencing; a query as to which of two methods of interpolation Hind used at the Almanac Office to sub-tabulate values (hourly values of the Moon's position, for example) from fixed known pivotal values (twelve-hourly positions); and finally, whether the availability of a machine would induce him to abandon a non-mechanisable method in favour of one amenable to machine computation.

Hind's responses are terse and unwaveringly negative. He wrote that he would prefer to employ a human computer with pen and paper rather than an infrequently used machine because 'very few errors are committed and the work is performed quicker than a machine would turn out'. Also that 'any practised computer, with the aid of Tables we now use, would beat such a machine as that constructed by Mr. Scheutz in a [... period] of four or five hours'.[68] The upshot of his testimony is that for the Nautical Almanac machines did not offer any advantage over human computers in respect of speed or accuracy.

The response from the GRO to Airy's survey was as determinedly positive as the Nautical Almanac's was negative. Farr argued in a long authoritative submission that the extra hands required to manually compute the tables for the expanded Life Table would be more costly than the machine and that manually computed tables would not enjoy the same confidence in their accuracy. He had already started the computations for joint lives using human computers and was able to quantify the labour involved.[69] He had hard data to back his case.

In his report to the Treasury, sent on 30 September 1857, Airy gave his verdict. Answering himself for the Greenwich Observatory he repeated the case he had made to Goulburn 15 years earlier:

> In the Royal Observatory, the Machine would be entirely useless ... During the twenty two years in which I have been connected with the Royal Observatory, not a single instance has occurred in which there was a need of such calculations.[70]

On behalf of the Nautical Almanac he reported:

> [The superintendent's] opinion is clear and unhesitating, that no advantage would be gained by the use of the Machine, and that he would prefer the pen-computation of human computers, in the way in which it has hitherto been employed.[71]

But Farr's advocacy was not so easy to dismiss and Airy, against his better judgement, agreed to the purchase of a copy of the machine. He conceded that there would be value in such an experiment should the machine prove to be of use, and there was anyway a great deal of public interest in mathematical machines.

The machine was completed in July 1859 by Donkin & Co. at a loss of £600. In the event the second Scheutz engine made only a slight contribution to the 1864 Life Table. The hoped-for benefits of mechanised tabulation were not realised: only 28 of the 600 pages were composed entirely by the machine with little financial saving, and reliability was a source of constant anxiety. Farr's campaign had failed, and Airy's consistent scepticism was seemingly vindicated.

But there is one last twist in the plot. Late in the day Airy underwent a partial conversion. Before signing off final payment for the engine to Donkin, Graham requested Airy to perform an acceptance inspection. This Airy duly did with some thoroughness on 31 August 1859. In the course of putting the machine through its paces Airy suddenly saw a use for it: producing tables in degrees, minutes, and seconds of arc for astronomical use rather than in the standard decimal formats preferred by both Babbage and Scheutz. Airy excitedly wrote to Donkin asking whether the machine could easily be converted from decimal to sexagesimal working and whether the printwheels could be modified to alter the formatting of results on the page. Donkin replied that the modifications required were minor – a matter of a few hours. Airy approved receipt of the machine commending its construction and enabling settlement with Donkin. Nothing came of his late burst of enthusiasm, and he bowed out taking no further interest in promoting the use of the machine. So ended the last of Airy's official consultations.

3.8 Not Just Engines

There were several instances of non-governmental consultation in which inventors and entrepreneurs solicited Airy's views on a variety of mechanical calculating aids. Notable amongst these are Thomas Fowler (1777–1843) and William Bell. Fowler devised and built a digital calculator using ternary arithmetic and sliding rods. It was made of wood and operated by hand.[72] Versions of the calculator were constructed in 1840 and 1842, and Fowler demonstrated his invention in London to some acclaim. Babbage, Francis Baily, and Augustus de Morgan among others witnessed it and commended its operation. Airy's views on the machine were invited by Fowler's patron Sir Trevor Wheler, and here Airy reveals new reservations about mechanised calculation in general. The basis of his opposition was not any inherent deficiency in calculating devices, but the expertise required by staff to derive any benefit from them, and the difficulty of educating staff in their use:

> The number of persons who can use even the common sliding rule is very small. I do not mention this as tending to exclude absolutely the advantage of such mechanical contrivances, but as tending to limit it presently. They will only be used when there is a systematic preparatory education: thus all my Assistants are instructed to use the sliding rule … without this express training the sliding rule would not be used. This applies more strongly to a more complicated machine especially when the precise [?] reduction is so heavy as it must be in Mr. Fowler's.[73]

The Astronomer Royal's response offers a sobering view on the practicalities of introducing even rudimentary aids in an institutional context for routine work (Figure 3.4).

In the case of William Bell's calculator Airy expressed a related set of reservations. Bell's calculator was a circular logarithm instrument capable of continuous division or multiplication. The principle was that of a circular slide rule but with a self-checking ability to keep track of the decimal point. Bell wrote to Airy in August 1849 seeking his endorsement for the instrument with a view to bringing the device to the attention of the Royal Society.[74] Airy affirmed that the instrument performed as Bell described and confirmed the desirability of tracking the decimal point, but swiftly followed this by saying that he was 'confident that the instrument will never be used'. His reasons, as with the Fowler machine, were based on an unfavourable comparison with his bench-mark of operational convenience and established use – the slide rule: Bell's instrument required too many steps to operate, was more difficult to use, and offered little improvement in accuracy.[75]

Airy was not the first to greet innovative calculating machines without enthusiasm. In 1648 Balthazar Gerbier, an Anglo-Dutch courtier, argued that Blaise Pascal's calculator was prohibitively expensive for all but the wealthy and that its operation anyway presupposed a good command of arithmetical practice. He advocated instead 'the ordinary reddy way' of established practice.[76] Robert Hooke was similarly dismissive of the practical utility of both Samuel Morland's calculator and Gottfried Leibniz's. Of the latter he wrote: 'I could not perceive it ever to be of any great use especially common use … [I]t could only be fitt for great persons to purchase … and

FIGURE 3.4 Mark Glusker and the reconstruction of Thomas Fowler's Ternary Calculator. (Courtesy of Mark Glusker.)

for great witts to understand and comprehend'.[77] Three centuries later Airy joined the chorus opposing the adoption of mechanical aids as a replacement for existing practice.

3.9 Airy's Case against Calculating Machines: Summary

With the exception of a late short-lived flare of enthusiasm for the altered use of the Scheutz engine, Airy remained consistently sceptical about the utility of automatic calculating engines and mechanical calculating devices. He rejected the notion that there was any need for calculating engines of the kind proposed by Babbage, arguing in consultative reports, publication, and private correspondence that manual methods of tabulation were sufficiently accurate, that the demand for new tables was non-existent or at best infrequent, and that there was no economic case to justify the high capital costs of complex automatic machines. He further argued that machines were anyway not the whole solution: even if they operated infallibly, the integrity of any results still depended on manually calculated starting values that were labour-intensive and required high levels of mathematical expertise. Finally, he maintained that the demand for calculating engines was led by entrepreneurs and technologists, not by the real computational needs of contemporary practice of which the engine advocates appeared ignorant. His views on mechanical calculating aids were equally negative. Here he argued that the operation of such devices was too complicated, that training for general use was impractical, and that they anyway offered no improvement in accuracy.

The universe of reference for Airy's assessments was the Greenwich Observatory, and the utility of machinery was throughout assessed against the needs of the routine tabulations undertaken there, with the humble slide rule and existing tables providing the bench-mark of operational convenience and efficacy. A new age of machine computation with implications for new branches of mathematics as envisaged by Babbage held no promise for the pragmatic Astronomer Royal. In the case of Neptune, we find a man unable to lift his gaze from his household accounts to gaze at the heavens to behold a new planet. When it came to the visionary potential of automatic computing machines, it seems that he placed greater value on ensuring that his clerks were competent in the use of the slide rule. In this he succeeded excellently well.

4

The Royal Observatory 1881–1998

Tony Mann

CONTENTS

4.1 Introduction

This chapter will bring to its conclusion the story of the Royal Observatory at Greenwich, covering the Astronomers Royal who followed Airy at Greenwich, the Observatory's moves to Herstmonceux and Cambridge, and its eventual closure.

4.2 The Eighth Astronomer Royal: Sir William Christie and his Time

George Biddell Airy, the subject of the previous chapter, was succeeded as Astronomer Royal by Sir William Henry Mahoney Christie (1845–1922) (Figure 4.1). Christie had a mathematical background: his father, Samuel Hunter Christie (1784–1865), was Second Wrangler in 1805 and had shared the Smith's Prize. The father subsequently taught mathematics at the Royal Military College, Woolwich, and was from 1837 to 1854 Secretary of the Royal Society. He worked on magnetism and the compass, and Cape Christie in the Antarctic was named after him.[1]

The future Astronomer Royal was Fourth Wrangler in 1869 and was recruited as chief assistant at the Royal Observatory in 1870 by Airy, who had known his father and who had been impressed by the son's performance in his unsuccessful examination for the Sheepshanks exhibition in 1866.[2] Christie studied the causes of systematic errors in observations, showing for example that small errors were resulting from wear to the micrometer screws in the reading microscopes, and he was largely responsible for the introduction of spectroscopy and photography at the Royal Observatory.[3] He was appointed eighth Astronomer Royal in 1881 on Airy's retirement (John Couch Adams having turned down the position).[4] During his time as Astronomer Royal

FIGURE 4.1   Sir William Henry Mahoney Christie (photograph by Elliot and Fry). (Reproduced from E. Walter Maunder, *The Royal Observatory Greenwich: A Glance at its History and Work* (London: The Religious Tract Society, 1900).)

he oversaw a number of additions and adaptations to the Observatory buildings and brought the Observatory into line with recent developments in astronomy.[5] A new 28-inch refracting telescope, for a time the largest in the world, was installed by 1894, with an object lens which, following a suggestion by Sir George Gabriel Stokes, could also be adjusted to be used for photography.[6] A new building housing a 26-inch refractor and a 30-inch reflector was completed in 1898. This South Building was designed to be completed in stages, since Christie knew that he would not initially have been granted funding for such a large building.[7] The telescopes were provided by Sir Henry Thompson (1820–1904), the surgeon who introduced the practice of cremation.[8]

Christie promoted international co-operation between observatories. The Paris conference of 1887 initiated a project, the *Carte du Ciel*, in which 18 observatories collaborated to produce a photographic map of the entire sky, and it was during his time as Astronomer Royal that the International Meridian Conference in Washington DC voted to make Greenwich the Prime Meridian.[9]

It is thankfully rare for a mathematical institution to be the subject of a terrorist attack. One of the most curious episodes in the history of the Royal Observatory occurred at eight minutes to five on the afternoon of Thursday 15 February 1894. Some schoolboys heard an explosion and, rushing to the path near the Observatory, found a seriously injured young man. This turned out to be Martial Bourdin (1867/8–1894), a French tailor living in London and an anarchist, who had been injured by the explosion of a bomb he had been carrying. Bourdin was taken to the Royal Naval Hospital where he died of his wounds within a few minutes. His motives remain unclear. The inquest was told that the bomb could not have been detonated accidentally, and the assumption was that he had been attempting to attack the Observatory, which was in fact undamaged apart from a chipped brick. Other anarchists said that a scientific institution would not have been targeted in this way, and the involvement of an *agent provocateur* was suspected by some. The episode, coming a few days after a fatal bomb attack on a café in Paris, caused a somewhat hysterical reaction, raising public concern about government asylum and immigration policies. Joseph Conrad, who told his publisher he had 'inside knowledge', used this incident as the basis for his short novel of 1907, *The Secret Agent*, in which a naïve young man blows himself up in Greenwich Park while intending to bomb the Observatory, the attack having been instigated by an *agent provocateur*. Curiously, in turn, this book may have had an effect upon Theodore Kaczynski, the former mathematician who conducted the lethal Unabomber bombing campaign in the United States between 1978 and 1995.[10]

In 1900, during Christie's time as Astronomer Royal, a richly illustrated 'short account of [the Observatory's] history, principal instruments, and work' was published.[11] Its author, Ernest Walter Maunder (1851–1928), had been appointed as a photographic and spectroscopic assistant at the Observatory in 1873, and worked there until his retirement in 1913, and again during the First World War. Maunder carried out work in spectroscopy with Christie. He was Secretary and then Vice-President of the Royal Astronomical Society, but was also founder of the British Astronomical Association, of which he was President from 1894 to 1896[12]: he 'wanted an association of astronomers open to every person interested in astronomy, from every class of society, and especially open for women'.[13]

Maunder's second wife was Annie Scott Dill Russell (1868–1947). She had been ranked as a Senior Optime (equivalent to second class honours) at Cambridge but, as a woman, was unable to receive a degree. After a year as a schoolteacher, she was appointed a 'Lady Assistant' at the Observatory in 1891. She had to resign her position when she married Maunder in 1895, but returned and worked at the Observatory as a volunteer between 1915 and 1920. She was elected a Fellow of the Royal Astronomical Society in 1916 after the Society changed its rules to allow women to become Fellows. Walter and Annie Maunder are noted for their work on sunspots, both together and independently. The period in the sixteenth century in which sunspots were rare is known as the 'Maunder minimum'.

Christie retired as Astronomer Royal in 1910 and died at sea in 1922. He was succeeded by Sir Frank Watson Dyson (1868–1939).

4.3 The Ninth Astronomer Royal: Sir Frank Dyson and his Contemporaries

Dyson (Figure 4.2) had been Second Wrangler in 1889 and Smith's Prizeman with his friend Hector Macdonald in 1891.[14] He played a leading role in the coup achieved by the Cambridge magazine *Granta* in 1892 when, within 24 hours of the Mathematical Tripos problem paper, it published a 'Mathematical Supplement' giving answers to all

FIGURE 4.2 Sir Frank Watson Dyson (photograph from Bain News Service). (George Grantham Bain Collection, Prints and Photographs Division, Library of Congress, LC-B2-5158-11.)

the questions set. Dyson and Macdonald had spent the time solving all the problems (with help from a student who had sat the paper and solved the one problem which had defeated the other two), and rushed their solutions to the printer as each one yielded.[15]

Dyson was appointed Chief Assistant at the Royal Observatory in 1894. He acquired a reputation for 'soundness' and as an effective collaborator: many of his publications were joint ones, reflecting the influence he had on astronomy both nationally and internationally.[16] He worked on the *Carte du Ciel*, Greenwich being the only observatory to complete its allotted share of the work by 1909, and with William Grasett Thackeray (1853–1935)[17], he applied modern error correction methods to the 27,000 observations gathered by Stephen Groombridge (1755–1832) at his private observatory in Blackheath between 1806 and 1816. These had been published after Groombridge's death by Airy in 1838, and the revised observations were particularly important at the time in the context of work by Jacobus Cornelius Kapteyn (1851–1922), who studied the Milky Way and was the first to find evidence for galactic rotation. Groombridge's data were used by Sir Arthur Eddington (1882–1944), who succeeded Dyson as Chief Assistant in 1906, in his work confirming Kapteyn's suggestion on stellar motions.[18]

In 1905 Dyson moved to Edinburgh as Astronomer Royal for Scotland, and on Christie's retirement in 1910 he returned to Greenwich as Astronomer Royal.

It was Dyson who realised in 1917 that the total solar eclipse of 29 May 1919 would provide an opportunity to test the prediction of Einstein's General Theory of Relativity that light would be deflected by the gravitational field of the sun. Despite the difficulties of planning during wartime, Dyson arranged for observations to be made by A.C.D. Crommelin (1865–1939)[19] and C.R. Davidson (1875–1970)[20] at Sobral in Brazil, and by Eddington and Edwin Turner Cottingham (1869–1940)[21] at Principe in West Africa. Although a present view is that the results were less than conclusive,[22] at the time Dyson (who according to Eddington 'had not wholly expected the result that was obtained'[23]) and Eddington interpreted the data as confirming Einstein's prediction, and famously after Dyson described the experiments at the Royal Society the Times headlined its report, 'Revolution in Science. New Theory of the Universe. Newtonian Ideas Overthrown'.[24]

Dyson also left his mark on British life through his introduction of the 'Greenwich time signal' – the six familiar 'pips', the final one marking the hour, which have been broadcast (principally for ships) since 1927 – another icon of Greenwich's contribution to navigation.

When, after the Great War, the electrification of the local railway service threatened to make the magnetic investigations at the Observatory impossible, Dyson, rather than resist the development, negotiated that the railway company should pay for the move of the magnetic work to Abinger in Surrey.[25]

Admiral of the Fleet the Earl of Cork and Orrery, W.H.D. Boyle (1873–1967), who was President of the Royal Naval College from 1929 to 1932, recalls a visit to the Royal Observatory:

> On the first visit which my wife and I paid to it we were taken over by the late Sir Frank Dyson, at that time the Astronomer Royal – a delightful personality. He showed us much of interest and then said: 'Now I will show you the clock from which the world takes its time,' and turning a corner he pointed dramatically to a clock … which had stopped! He had of course missed his way and we were in front of the wrong timepiece![26]

In 1920 an Admiralty hydrographer, Rupert Gould (1890–1948), who was working on a history of the marine chronometer in his spare time, discovered in the Observatory cellars the timepieces made by John Harrison in his ultimately successful attempt to win the Longitude Prize. Dyson gave Gould permission to restore them at his home. When Gould lost his job in 1927, following a sensational court case when his wife sought judicial separation, he devoted himself to the clocks: the restoration was completed in 1935, and they are now proudly displayed in the Observatory. Gould, who also wrote a biography of Captain Cook, was subsequently employed by the BBC as the Stargazer, and found fame as a panellist and adversary of Professor Joad on the radio programme *Brains Trust*.[27]

Observing eclipses is very dependent on the weather conditions at the time, and Dyson was reputedly remarkably lucky in that respect: visibility was good for all six of the eclipses that he observed personally, and he described himself as 'a hundred percent eclipse observer'. But according to one anecdote he may have been less than tidy: when his daughter asked him what was meant by 'chaos', he replied, 'Chaos? chaos? well – go and look at my writing desk'.[28]

It seems that Dyson did not get on well with Philip Herbert Cowell (1870–1949), the Superintendent of the Nautical Almanac Office. Cowell had applied for both the Plumian Chair in Astronomy and the Lowndean Chair in Astronomy and Geometry at Cambridge when they fell vacant in 1912 and 1913, but Eddington and H.F. Baker, respectively, were appointed, and Cowell felt that Dyson had not supported his candidacies. After this Cowell effectively abandoned his own research.[29] Previously in 1909 he and Crommelin had calculated the time of perihelion for the 1910 visit of Halley's comet: in the event their prediction was only 2.7 days early. Cowell and Crommelin, who received the Lindemann Prize of the Astronomische Gesellschaft for this work, suggested that non-gravitational forces might be responsible for this difference, which was confirmed to be the case in 1986.[30] H.H. Turner, whose election to the Savilian Chair of Astronomy at Oxford had created the vacancy as Chief Assistant when Dyson was appointed, celebrated the comet in song

> Of all the meteors in the sky
>   There's none like Comet Halley,
> We see it with the naked eye
>   And periodically.
> The first to see it was not he,
>   But still we call it Halley,
> The notion that it would return
>   Was his originally.

which concluded after many more verses

> They say the greatest man is not
>   A hero to his valet,
> But I'd rejoice if 'twere my lot
>   To serve the fame of Halley.
> So Cowell and Crommelin, Davidson,

> Knox-Shaw,[31] and all the tally,
> We'll toast the splendid work they've done
> Enthusiastically.[32]

His obituary recalls Cowell's retirement:

> On attaining the age of sixty in 1930, he arranged that his nephew should be waiting with a car on his birthday, at the precise hour of his birth, outside the Nautical Almanac office: as he emerged for the last time, hatless and with the light on his thick white hair, he said on approaching the car 'Old Kaspar's work is done'.[33]

A number of other notable astronomers worked at the Observatory under Dyson. Davidson has already been mentioned, as has Crommelin (who was President of the Royal Astronomical Society 1920–1930), and there will be more of Harold Spencer Jones shortly. Sydney Chapman (1888–1970) was Senior Assistant from 1910 to 1914 and also worked there during the war, having declined, as a pacifist, to do scientific war work: he subsequently held posts at Trinity College, Cambridge, Manchester University, Imperial College, London and The Queen's College, Oxford. Between 1912 and 1919 he found exact solutions to the problems of gas viscosity, heat conduction, and diffusion, posed by Maxwell and Boltzmann. Amongst many other such distinctions he was President of the London Mathematical Society (1929–1931), the Mathematical Association (1932–1934), the Royal Meteorological Society (1932–1933), and the Royal Astronomical Society (1941–1943), and received the Royal Society Royal and Copley Medals, the LMS de Morgan Medal and Larmor Prize, and the Gold Medal of the Royal Astronomical Society.[34]

On Chapman's resignation John Jackson (1887–1958) was appointed Chief Assistant. The dangers that observation sometimes involved were shown during the war when a Zeppelin dropped a bomb within the Observatory grounds: Jackson carried on observing. He worked on the theory of double stars, and on the accuracy of clocks, showing that the Shortt free-pendulum clock, installed at the Observatory in 1925, while more regular than apparent sidereal time, which is affected by the nutation of the earth, was not sufficiently accurate to detect any irregularities in the earth's rate of rotation. In 1933 he left Greenwich to work at the Royal Observatory at the Cape of Good Hope.[35]

Michael Greaves (1897–1955) was a Cambridge Wrangler in 1919 and was awarded the Isaac Newton Fellowship. He benefited from an arrangement instigated by Sir Joseph Larmor whereby Isaac Newton Students worked for a few months at the Royal Observatory (Leslie Comrie, the subject of Chapter 9, was the only other to have this opportunity). Greaves was subsequently appointed second Chief Assistant in 1924 and worked with Davidson on the mathematics of the determination of effective stellar temperature. In 1938 he became Astronomer Royal for Scotland. After the Second World War he and L.S.T. Symms showed that the Post Office's quartz crystal clocks were superior to the Shortt free-pendulum ones. He was President of the Royal Astronomical Society from 1947 to 1949.[36]

Dyson retired in 1933. In 1939 he, like his predecessor Christie, died and was buried, at sea. His successor as tenth Astronomer Royal was Sir Harold Spencer Jones (1890–1960).

4.4 The Tenth Astronomer Royal:
Sir Harold Spencer Jones

Spencer Jones had been a Wrangler at Cambridge in 1911, Isaac Newton Student in 1912 and second Smith's Prizeman in 1913, in which year he became Chief Assistant at the Royal Observatory. In 1923 he was appointed Astronomer Royal at the Cape of Good Hope, where he worked on the determination of an improved value of the solar parallax: this work gained him the Gold Medal of the Royal Astronomical Society and a Royal Medal of the Royal Society, both awarded in 1943.[37] He was to hold all the offices of the Royal Astronomical Society, including the Presidency from 1937 to 1939.

Spencer Jones succeeded Dyson in 1933 and continued Dyson's work on time distribution. In 1936 he introduced the Speaking Clock (in collaboration with the Post Office).[38] But time itself was running out for the Observatory at Greenwich, where, now that it was one of the suburbs of London, street lighting and smoke were making astronomical observations impossible. Spencer Jones realised that the Observatory would have to move, and a suitable site was identified at Herstmonceux Castle in Sussex.

The move was delayed by the Second World War, during which the big telescopes were removed from Greenwich for safe keeping: indeed a bomb did damage the telescope building. After the war the site at Herstmonceux was acquired and the move began, being finally completed in 1957. At this time the Royal Observatory was renamed the Royal Greenwich Observatory.[39] The big telescopes were fully working at Herstmonceux by the end of 1958, three years after Spencer Jones had retired.

4.5 The Eleventh Astronomer Royal:
Sir Richard van der Riet Woolley

In 1933 Richard van der Riet Woolley (1906–1986) succeeded Jackson as Chief Assistant.[40] Woolley was born in Dorset; his family moved to Cape Town in 1921. He was a Wrangler in 1928 and also had first class BSc and MSc degrees from Cape Town University. The comment that 'On paper Woolley's academic record was perhaps less outstanding than that of any of the eight or so then living predecessors' on appointment as Chief Assistant demonstrates the mathematical talent regularly available to the Observatory and its attractiveness to brilliant young mathematical astronomers.[41]

Woolley returned to Cambridge as John Couch Adams Astronomer in 1937. He was appointed the eleventh Astronomer Royal at the beginning of 1956, where he oversaw the completion of the move away from Greenwich. The site at Greenwich passed to the National Maritime Museum, and in 1960 the Queen opened Flamsteed house as part of the Museum.

4.6 The End of the Royal Greenwich Observatory

The Royal Greenwich Observatory moved again in 1990, from Herstmonceux to Cambridge, where it shared facilities with the Institute of Astronomy, but eight years later, 323 years after its foundation by Charles II, the Observatory was finally closed by the Particle Physics and Astronomy Research Council, with some of its functions transferring to a new Astronomy Technology Centre in Edinburgh, while the Nautical Almanac Office moved to the Rutherford Appleton Laboratory in Didcot, Oxfordshire.[42] Its historic buildings at Greenwich are now open to the public under the auspices of Royal Museums Greenwich.

5

Mathematics Education at the Greenwich Royal Hospital School

Bernard de Neumann

CONTENTS

5.1 Introduction

This chapter concentrates upon the evolution of an early institution that was based in Greenwich, and which made, through its teachers and pupils, a very great contribution to the establishment of Britain's Empire and Sea Power: Just as, according to the Duke of Wellington, the Battle of Waterloo was won on the playing fields of Eton, it may justifiably be claimed that the establishment, defence, integration, and trade of the largest empire the world has yet seen, the British Empire, was charted and plotted in the classrooms of the Greenwich Royal Hospital School and facilitated by its former pupils.

5.2 The Royal Hospital School

After the repulsion and defeat of the Spanish Armada of 1588, some influential sailors (notably John Hawkins and Francis Drake) set about establishing a fund (The Chatham Chest) for the welfare of injured and aged seamen. In part this action was motivated by the disgraceful treatment of the British sailors who had taken part and who following their successful defeat of the Spanish were left unpaid and languishing in their ships. Typically of such ventures of those times, the funds were fraudulently (by today's standards) usurped into the hands of outsiders, who often lived extremely comfortable lives at the expense of the contributing sailors. Nevertheless a seed of an idea had been planted.

Following the establishment of the Army's Royal Hospital at Chelsea in 1682 (which itself was modelled upon the Royal Hospital, Kilmainham, Dublin, begun in 1680 for the wounded and infirm from the Royal Irish Army), the British monarchy in the person of Queen Mary became concerned about the welfare of British sailors who often defended the British Isles from the unwelcome attentions of continental powers. Accordingly, much consultation and discussion took place, and eventually a Royal Charter was drawn up to honour her wishes, which had become 'the darling object of her life' before she died. Mary died before the Charter received royal assent, and so King William as a mark of his affection for his newly deceased Queen had it post-humously dated so that Mary's great concern and contribution to the scheme should be properly remembered. Thus, with the date 25 October 1694 the Royal Charter of Greenwich Hospital appeared which planned to[1]

> erect and found an Hospital within Our Mannor of East Greenwich in Our County of Kent for the reliefe and support of Seamen serving on board the Shipps or Vessells belonging to the Navy Royall . . . who by reason of Age, Wounds or other disabilities shall be unable to maintain themselves. And for the Sustentation of the Widows and the *Maintenance and Education of the Children of Seamen* happening to be slain or disabled . . . Also for the further reliefe and Encouragement of Seamen and *Improvement of Navigation*.

The educational aims of the Royal Charter seem to have been added, almost as an afterthought, in order to engage the Royal Mathematical School (founded by Charles II in 1673) of Christ's Hospital in some competition, encouraging it to improve its standards, thereby producing high standards from both schools. The Royal Mathematical School had got off to a poor start as, after its inception, there was no demand from the Navy Royal for its King's Letter Boys. This led to its 'mathemats', who were generally bigger and stronger than their 'Grecian' counterparts, becoming loutish, with much intimidation of their smaller brethren. It was not until after William Wales took up his position there in 1775 that these problems were finally eradicated and the school began to function properly. William Wales (1734–1798) contributed to astronomy through his observations of the 1769 transit of Venus, and his studies of latitude and longitude on Captain James Cook's second voyage to the South Seas. During this voyage, Wales was responsible for monitoring the performance of the chronometers. After returning from the Pacific, Wales took charge of the Royal Mathematical School at Christ's Hospital in London, and, over the next two decades, he taught a succession of budding officers the principles of navigation. He also achieved iconic status at the school becoming 'Grecian' Samuel Taylor Coleridge's model for the 'Ancient Mariner'! Another school founded to compensate for the inadequacies of the Royal Mathematical School's performance was Sir Joseph Williamson's Mathematical School, Rochester, which was founded in 1701.

In 1696 in the Act for the Increase and Encouragement of Seamen, the benefits of Royal Hospital Greenwich were extended to include 'such mariners, watermen, seamen, fishermen, lightermen, bargemen, and keelmen as shall voluntarily come in and register themselves in and for His Majesty's sea service'. It should be recognised that in those days, sailors were simply sailors with no particular allegiance to the Crown. The 'Navy Royall' consisted of a few ships and other vessels that notionally belonged to the Crown and were used for the Monarch's purposes – including war. In times of war, merchant vessels were also chartered, together with their crews, and equipped with arms and soldiers, under the command typically of an Army captain and subordinate lieutenant. Ships were under the command of a Master, who was responsible for all aspects of his ship's operation, including the welfare of his crew. The ship's owner appointed a Master for a voyage or series of voyages, and he then selected the crew who then 'signed on' by signing the ship's agreement which bound them to follow the lawful command of the Master. Masters' duties also included keeping a detailed log of the voyage, noting wind, weather, and ocean currents at various times of the day, hazards encountered (including uncharted ones), and sketches of significant features like islands, headlands, estuaries, etc. (together with the direction from which they were viewed, and of course the ship's position). The information from such logs was collated by the Hydrographic Service to produce increasingly accurate navigation charts. All seamen (as we have already explained, there was no real distinction between royal and merchant seamen) were required to pay 6d per month into the Greenwich Hospital fund.

The basic design of the hospital was begun by Sir Christopher Wren, who himself possessed noted mathematical abilities. Subsequently, Wren saw to it that the foundations of the buildings were installed before he retired, and so later architects like Hawksmoor were very much constrained to follow Wren's basic floor plan. Pensioners began to take up occupation of their new quarters in January 1705, whilst work continued on constructing the remaining planned buildings. The number of pensioners accommodated eventually rose to over 2,500 before the Hospital went into decline and pensioners opted to live out. The Hospital closed its doors to In-pensioners in 1869, and the buildings were leased to the Royal Navy as a college in 1873.

In 1712, Greenwich Hospital felt that it could commence the education of children of seamen and began this in a small way by supporting Thomas Weston, assistant to Astronomer Royal Flamsteed, in his venture to begin an Academy in Greenwich. Flamsteed had been a tutor at the Royal Mathematical School at Christ's Hospital, and when he left to become Astronomer Royal, he took many of his pupils with him. Indeed, Weston had been a pupil of Flamsteed at the Royal Mathematical School. Thus began the teaching of mathematics in Greenwich.

The first Greenwich Hospital pupils were thus sent to Weston's Academy in Greenwich in 1712, where they received an excellent and highly valued education. In Weston's Academy at this time, the boys were accommodated in the attic of the King Charles block. Soon the number of Greenwich Hospital pupils grew to such an extent that it became economical to provide their own school and teachers. This school fully occupied the site of Weston's Academy, which itself moved to a new site, and eventually became Burney's Academy, led by William Burney, the brother of the novelist Fanny Burney. The Greenwich Hospital School became a great success through its teaching of mathematics, navigation, and nautical astronomy, providing its pupils with sufficient knowledge for them to become navigators and ships' officers in the Royal and Merchant Navies where they joined directly as Masters' Mates.[2]

The boys would have been taught to use such navigational instruments as the magnetic compass, Nocturnal, Back-staff, Cross-staff, Quadrant, and Sextant, and would have been familiar with map projections such as Mercator. They would have participated fully in the propagation of techniques for determining longitude, especially after Harrison's construction and Cook's demonstration of the sea-going chronometer. Such education was highly prized, but extremely unusual, in those times when hardly any mathematics was taught in schools, let alone universities, and grammar schools concentrated upon the 'practicalities' of speaking the dead language Latin!

Places in the school were much sought after and valued, and admittance to the school was restricted to the sons of officers and men of Navy Royal, Marines, and the merchant service. Candidates had to seek patronage from at least one of (i) the Lords of the Admiralty, (ii) the First Secretary of the Admiralty, (iii) the Governor and Lieutenant Governor of Greenwich Hospital, (iv) five Commissioners of the Hospital, and (v) the Chairman of Lloyd's Patriotic Fund. These exercised their patronage in rotation as vacancies arose. Naturally, there was occasional abuse of powers.

The rules of the school stated that no boy was to be admitted before the age of 14, nor retained after the age of 18, and furthermore that boys were 'To be put out as apprentices to Masters of ships and substantial Commanders, for better improvements of their talents, and becoming Able Seamen and Good Artists': the latter term in its original context and usage means 'skilled seamen and good navigators'. The school's leavers were much sought after by ships' captains, who often expressed amazement that boys so young were demonstrably competent to navigate the oceans. The school also provided the bulk of officers to the Hydrographic Service (the branch responsible for surveying the oceans of the world). The first person to pass the Extra Master's examination of Trinity House (the body responsible at the time for examining ships' officers) was a former pupil of the school. Long ago, the school acquired the title *Cradle of the Navy*, a title that may well derive from James Moncrieff's 1759 pamphlet *Three Dialogues on the Navy* in which it is stated in regard to Greenwich Royal Hospital: 'From the couch and sepulchre of age, I would change it into the cradle and, as it were, the forge of youthful merit'.

In 1798 in an independent development, The British National Endeavour, a boarding school in Paddington, was established, with fervent public support and subscription, following public concern at the loss of life and injuries sustained by British seamen during recent battles. The school rapidly outgrew its premises on Paddington Green, which could only take 70 children. Originally, it was a small 'industrial school' for children whose fathers had been seamen in the Royal Navy and had fallen in action. Lord Nelson was an early patron, as were the financiers, the brothers Abraham and Benjamin Goldsmid. Following the news that the French and Spanish had been defeated at Trafalgar, it was renamed under a Royal Warrant backdated to 21 October 1805 as The Royal Naval Asylum. The Asylum acquired the Queen's House and its estate, and moved to its new home in Greenwich. The new facilities (presently occupied by The National Maritime Museum) were formally opened on 21 October 1807. Only the children (boys and girls, orphans or motherless, or the father disabled or serving on a distant station) of sailors and marines were admitted to the Asylum aged between 5 and 12 and presumably left at normal school leaving age (14). Because of a generous donation valued at £61,000 by Lloyd's Patriotic Fund, Lloyd's were permitted to nominate children from other seafaring backgrounds for attendance at the Asylum. The Royal Navy had 'first refusal' on all boys leaving the Royal Naval Asylum. Rather loosely and age-wise, the Asylum corresponded to today's primary schools, whilst the Greenwich Royal Hospital School corresponded to a secondary school; however, their purposes were much different as most of the Greenwich Royal Hospital School's pupils were destined for service at sea.

The two schools operated independently side-by-side at Greenwich until 1821 when it was realised that it would reduce administrative overheads to merge the schools and their sites as, eventually, the Greenwich Royal Hospital Schools. In the merger, the old Greenwich Hospital School became the Upper School and the Asylum the

Lower School. In 1841, the Upper School was increased in size at the expense of the Lower School, and a third even higher school was established called the Nautical School, principally for teaching navigation and nautical astronomy. Under the guidance of Edward Riddle, and in succession his son John, the Nautical School became the globally acknowledged leading school in the instruction of navigators.

5.3 Navigation and Nautical Astronomy

John Seller, hydrographer to King William, wrote during the late seventeenth century that[3]

> ...the art of Navigation, is that part which guides the ship in her Course through the Immense Ocean to any part of the known World; which cannot be done unless it be determined in which place the Ship is at all times, both in respect of Latitude and Longitude; this being the principal care of a navigator and the Masterpiece of Nautical Science.

He added that

> four things are subordinate Requisites. Viz:
> Arithmetic
> Geometry
> Trigonometry
> The Doctrine of the Spheres [spherical, or mathematical, astronomy]

The entities that are dealt with in navigation and nautical astronomy are what we today call mathematical models. For example, mankind's idealised representation of the Earth has evolved from plane to sphere, to oblate spheroid, to geoid (earth-shaped), to paraphrase an old joke. In all cases, it is an idealisation, that may be specified relatively simply in mathematical terms, and so is a mathematical model.

Time, which is sensed by humankind, was originally understood and measured in terms of the apparent motion of the sun, moon, wandering stars (planets), and firmament. Complicated models of these motions were constructed in terms of geocentric interlocking spheres, which were later superseded by motions predicted by Newton's laws – again these succumbed in the end to more general, but nevertheless mathematical, models. Furthermore, prior to the discovery of quantum mechanics, and of mathematically chaotic solutions to seemingly deterministic dynamical systems, and the existence of 'strange attractors', and 'butterfly effects', the entire universe seemed deterministic and entirely predictable in the Leibnitzian sense. Thus, time could be scaled, measured, and determined by observing and measuring the apparent motions and relative positions of identifiable heavenly bodies. Hence, a navigator equipped with details of the predicted relative positions of certain easily identified, deliberately selected, celestial bodies, could, with the aid of instruments, calculate his position on the surface of the (idealised) Earth and further, with a knowledge of his destination, plan his route.

For the 'old school' mathematician, then the problem is solved and there is no further mathematical interest. However, navigators themselves were generally not mathematicians, yet they needed to execute faultlessly quite lengthy, some might be tempted to say tedious, computations once they had obtained and recorded their measurements. Furthermore, these computations were often carried out in testing environments, in a cramped, wet and windy, pitching, rolling, yawing, heaving, surging, and swaying cabin, lit by a candle or oil-lamp. Moreover, the only aids to computation were mathematical tables of various functions and the tables of stellar positions given in nautical almanacs.

Such activities generally necessitated interpolation, or inverse interpolation, for every look-up, and so, whilst such tables aided computation, the minimisation of their use was a helpful goal insofar as it helped to speed computations and increase the reliability of the results (tabulatory errors[4] and deliberately seeded inaccuracies[5] notwithstanding). Thus, a rudimentary precursor of computational complexity grew up, which sought to count and minimise the number of tabular look-ups required for each proposed method.

Efforts were also made to reduce the number of mathematical tables required as it was obviously impractical to carry tomes of tables, especially in small boats. Consideration was also given to the number of 'figures' these tables should be accurate to, and this at the time was set at six. Generally, a navigator carried his personal and trusted instruments, tables, and references from ship to ship, and this ability to minimise both computational effort and weight carried could be a real life saver before the days of radio, and Lieutenant Bligh's (of HMS *Bounty*), for example, or Shackleton's (of *Endurance*) desperate voyages from the Antarctic to Elephant Island and then on to South Georgia, attest to this. Thus, a navigator could manage with a sextant, a set of appropriate mathematical tables, a nautical almanac, and a compass, and knowing the potential destination. With none of these, it was generally only possible to steer in approximate directions, like the eight principal compass points.

Much effort was expended on attempting to minimise the 'computational complexity', thereby enhancing the likelihood of reliable results but also attempting to minimise the volume of tables required. For example, because of the many relationships that exist between the trigonometric functions, there is much scope for transforming the requisite navigational relationships, and some of these are, in principle, easier to compute than others.

Thus, in those days a navigator's 'toolkit' comprised a set of expressions that were amenable to calculation, a set of mathematical tables that, at a minimum, allowed the evaluation of these expressions with the minimum of effort/stress, and instruments and charts.

Furthermore, the effects of parallax, refraction of the atmosphere, dip to the horizon, and the sun and moon's apparent diameters, precession, and nutation, of the celestial poles were also modelled mathematically, so that their effects upon actual angular measurements could be compensated for by further calculations.

5.4 The Mathematical Riddles of Greenwich

Throughout the remainder of this chapter, we shall be concerned primarily with the Upper Nautical School of Greenwich Hospital Schools and, in particular, the influence of the mathematics teachers Edward and John Riddle.[6]

In October 1821, Edward Riddle was appointed Senior Mathematician at the Upper School. He was born into a poor family in Northumberland in 1786, and his general intelligence soon brought him to attend a school run by Cuthbert Atkinson in West Woodburn. Riddle and Atkinson's son (Henry, who married Riddle's sister, Isabella) both received an excellent education at this school, and both went on to establish their own schools. At the age of 18, Riddle became a schoolmaster and opened his own school shortly after in Otterburn, where his great enthusiasm for higher mathematics, astronomy, optics, and navigation knew no bounds. The *Ladies Magazine*, edited by Dr Charles Hutton,[7] in those times published short papers and letters on mathematical topics, and Riddle won prestigious prizes in 1814 and 1819. At the suggestion of Hutton, Riddle became Master of Newcastle's Trinity House School, and soon his fame spread far and wide. He probably journeyed by sea to London during this time and noted the deficiencies in the methods used by the colliers in navigating the North Sea. Again with the recommendation of Hutton, he was appointed a master at Greenwich.

Riddle was erudite by any standard,[8] and a great proponent and enthusiast for his subjects and pupils. It was his contention, contrary to most other teachers of navigation, that pupils would benefit greatly from a proper and thorough development of their subject which included concise and simple rules, with their investigations, for finding the latitude and longitude, and the variations of the compass, by celestial observations; the solution of other useful nautical problems, with an extensive series of examples for exercise.

He thus advocated real demonstrable understanding rather than learning by rote as was followed by the students of such as J.W. Norie.[9] Through his inspirational teaching, Riddle kindled a like enthusiasm in his pupils.

His most valued work was a *Treatise on Navigation and Nautical Astronomy* (first edition 1824) (Figure 5.1), which provided a complete course of mathematics for sailors, combining theory and practice in a way which had not been attempted before. It became

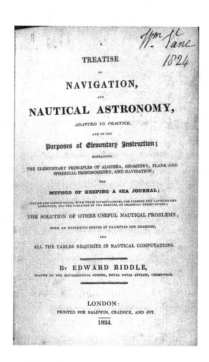

FIGURE 5.1   The title page of Riddle's *Treatise on Navigation and Nautical Astronomy*.

one of the standard textbooks for ships' masters and went through numerous editions, the eighth and final one appearing in 1864. It is worthy of note that the first edition was written in essentially three parts: Elementary Principles of Relevant Mathematics, Navigation, and Nautical Astronomy, with short pieces on the effects of winds and tides. In later editions, the Elementary Principles part was dropped, presumably reflecting the availability of other elementary mathematics books, and the treatise now had

Part 1: Navigation

Part 2: Nautical Astronomy

Part 3: Investigation of the Rules in Navigation and Nautical Astronomy

and a fourth part:

Notes (which gave further details of some points and some data) and Questions (for solution).

Additionally, he re-edited Charles Hutton's *Mathematical Recreations* in 1840 and 1854. Between 1818 and 1847, he also published 16 papers on astronomical subjects, of which eight were in the *Philosophical Magazine*, five in *Memoirs of the Royal Astronomical Society*, and three in *Monthly Notices of the Royal Astronomical Society*. His most important papers were on nautical astronomy and on chronometers. In 1821, the *Philosophical Magazine* published Riddle's article 'On the present state of nautical astronomy', and in 1829, the *Memoirs of the Royal Astronomical Society* included his piece on finding the rates of timekeepers, which was notable for showing how the going rates of chronometers could be determined without using a transit instrument.

He even found time to write the mathematical portions of *The London Encyclopaedia*. During 1841, at the request of the Admiralty, Riddle's teaching techniques were scrutinised by Augustus De Morgan and Thomas Hall (professors of mathematics at University College and King's College, respectively), who rapidly reached the conclusion that they were indeed remarkably effective.[10] This investigation arose through the scurrilous insinuations of two masters appointed to the schools on the recommendation of presumptive education-pioneer Sir James Key-Shuttleworth. These two masters heavily criticised Riddle's teaching techniques and results, but were easily refuted by Riddle with the glowing report of the two professors.[11] Hall and De Morgan reported that the boys had answered all the questions with accuracy and enthusiasm and many, even from among the most ordinary, had the making of fine navigators. One of Riddle's habits was to awaken his class in the middle of the night if there was some celestial body, or event, that could be clearly viewed, and by this means, he kindled their astronomical curiosity. Riddle planned and gained approval for the school to have an observatory, and this was completed before he retired. Astronomer Henry Lawson (1774–1855) presented his eleven-foot refractor telescope, which was considered by Dollond to be the finest he ever made, to the school in 1851, and it was duly installed in the new observatory.

Riddle was considered by his contemporaries to be one of the leading teachers of nautical science, and after his retirement, many of his friends and former pupils expressed their very great appreciation by presenting him with his bust in marble, by the celebrated sculptor William Theed (Figure 5.2). The bust was presented to the school by Edward's grandson and was prominently displayed in the hall of the Queen's House before the school moved from Greenwich. It is now in the Devonport Mausoleum next to the National Maritime Museum. At one time, every officer in

ERECTED BY OFFICERS
OF THE ROYAL NAVY AND MERCANTILE MARINE
AND MASTERS OF GREENWICH HOSPITAL
AND OTHER NAVIGATION SCHOOLS, ALL OLD PUPILS,
TO PERPETUATE THE MEMORY
OF
EDWARD RIDDLE, ESQ., F.R.A.S.
AND OF
JOHN RIDDLE, ESQ., F.R.A.S.
– FATHER AND SON –
EACH IN SUCCESSION HEAD MASTER
OF THE NAUTICAL SCHOOL
OF THE GREENWICH HOSPITAL UPPER SCHOOL
AND
IN GRATEFUL RECOGNITION OF THE CONSTANT
AND EARNEST EFFORTS OF THESE DISTINGUISHED
MASTERS
TO APPLY THEIR HIGH MATHEMATICAL AND
SCIENTIFIC ATTAINMENTS AND THEIR GREAT
TEACHING POWERS TO THE ADVANCE OF
NAVIGATION
AND THE SUCCESS OF THEIR PUPILS.

THE SCHOOL DURING THE YEARS OF THEIR
MASTERSHIPS
(1821 – 62) BECAME THE FOREMOST IN THE KINGDOM
FOR INSTRUCTION IN NAUTICAL SCIENCE.

THE BUST OF MR EDWARD RIDDLE
WAS PRESENTED TO HIM BY OLD PUPILS IN 1850
AND HIS GRANDSON MR E. RIDDLE, M.I.E.E.
GAVE IT TO TJE SCHOOL IN 1891.

FIGURE 5.2 The 'Classical-style' bust of Edward Riddle, now in Devonport Mausoleum in Greenwich, and the text on the monument.

the Hydrographic Service except one was a former pupil of Riddle, and many of his pupils went on to found and head Board of Trade Schools of Navigation. Others, with highly honed calculation skills, not wishing to go to sea, were employed as computers by the Royal Observatory or Nautical Almanac. A truly remarkable legacy.

Edward Riddle retired in 1851, and the Governors of Greenwich Hospital granted him a pension equal to his full pay, and he died in 1854.

During Riddle's tenure, in 1834, the Rev George Fisher joined the school as Chaplain and Master of the Upper School. Fisher had previously been a chaplain in the Royal Navy and taken part in the 1818 Admiralty Arctic expedition. He was the target of some criticism for neglecting his spiritual duties in 1840, and in 1841, the girls' school finally closed, and Riddle was appointed Master of the Upper School in place of Fisher who retained his position as Chaplain. The Mastership of the Lower School was vacant, and with Key-Shuttleworth's intervention, Riddle became Master of the Nautical School (a new higher school to instruct 200 of the ablest pupils), with two newcomers, William Graham and Thomas Irvine becoming Masters of the Upper and Lower Schools, respectively. Whilst at the school, Fisher wrote textbooks on algebra and geometry to introduce modern methods into the school and he superintended its astronomical observatory, observing there the solar eclipse of 18 July 1860, the year in which he acquired the new title of Principal Master of the Royal Hospital Schools. His books led to Riddle dropping his coverage of these mathematical aspects from his book. Much of his work was statistical in nature, which he applied to such problems as measuring the effect of iron on chronometers.

An important innovation of Fisher's was to evaluate the attainment and progress of his pupils numerically, and his system, with some modification and extensions, remains in use today throughout the world.[12]

John Riddle was born at Newcastle-upon-Tyne in November 1816 and was the only son of Edward Riddle. He was educated by his father and at the early age of 15 was appointed an Assistant Master in the Greenwich Hospital Schools and on the retirement of his father in 1851 was chosen to succeed him as Master. In 1846, he was elected to a Fellowship of the Royal Astronomical Society and from that time until his death was a regular participant at its meetings and an occasional contributor of papers. In 1854, he was appointed Examiner in Navigation to the Science and Art Department of the Committee of Council on Education and held a similar appointment in the Society of Arts, giving valuable service to both. As a teacher, Mr. John Riddle was probably unrivalled, and his success in teaching his young pupils not only the practice but also the theory of Navigation was remarkable (Figure 5.3). The influence he, like his father, possessed with his pupils was unbounded, as was his enthusiasm for his subject; a large number of scientific officers of the Royal Navy were greatly influenced by his vigorous mind and rigorous methods. He died as the result of an accidental fall from a platform in his classroom whilst teaching on the 11 October 1862, tragically only eight years after his father, and his untimely death was a great loss to the school and the Royal Navy.

Both Edward and John Riddle are still revered by pupils of The Royal Hospital School, where their achievements are remembered.

FIGURE 5.3 This photograph taken in 1855 is one of the earliest of a classroom. It shows John Riddle and his class at Greenwich. One row back, standing to the right is Albert Escott. (Courtesy of the Royal Hospital School, Holbrook.)

Albert Escott a former pupil of Edward Riddle, and an Assistant Master, was appointed in place of John Riddle, and he continued to teach in the Riddle tradition using Edward Riddle's book.

Then, in 1863, the grade of Master in the Navy was replaced by that of navigation lieutenant and that of Master's Mate by navigation sub-lieutenant, recruited not from the school but from H.M.S. *Britannia* (a Royal Naval cadet training ship moored at Dartmouth from 1869). This greatly reduced the potential for the school's prospective navigators, and this was further reduced by the Board of Trade regulations regarding the education of mercantile marine officers. The school, at this setback, continued to teach seamanship, navigation, and nautical astronomy to its pupils together with topics more generally taught in schools, but it marked the beginning of the demise of the Upper Nautical School. Gradually, the school introduced such subjects as thermo-dynamics, principles of steam-engineering, and electricity. This enabled its former pupils to become ships' engineer-officers, and again they were much sought after for service in merchant and Royal Navy vessels, until engineer training and education became much more formalised.

Over succeeding years, the naval emphasis of the school's education was gradually diminished and essentially ceased just before the Second World War. The Royal Hospital School continues to flourish. It moved to Holbrook, Suffolk, in 1933 and is now a regular boarding school, still bent on its naval traditions, but educating its pupils to fill useful places in modern society. The remarkably prescient vision of Queen Mary in promoting what she was pleased to refer to as 'the darling object of her life', has throughout three centuries borne fruit in many corners of the globe. For example, the first Governor of Australia, Admiral Arthur Philip, was a pupil; the Canadian Coast Guard College in Sydney, Nova Scotia, was originally modelled on the school; and many other nautical colleges, academies, and schools were strongly influenced by, or modelled on, Greenwich Royal Hospital Schools. Other international links include a sister institution also founded by its original Royal patrons: William and Mary College, Williamsburg, Virginia, USA.

This chapter has shown how many of the teachers, particularly the Riddles, of The Greenwich Royal Hospital School mathematics staff, together with their former pupils, fashioned much of the world as we see it today, and helped facilitate and defend trade on a world-wide scale.

6

The Royal Naval College

Tony Mann

CONTENTS

6.1 Introduction

This chapter will discuss the mathematicians who taught at the Royal Naval College from its arrival at Greenwich in 1873 until it, like the Royal Observatory, ceased to exist at Greenwich.

6.2 The Royal Naval College Moves to Greenwich

The move of the Royal Naval College from Portsmouth to Greenwich, into the former Greenwich Hospital where it opened in February 1873, had owed much to the desire of the Greenwich MP, William Ewart Gladstone, to provide jobs in his constituency following the closure of the Greenwich Hospital and of local dockyards. (This was sufficiently well known to be stated as a fact by the mathematical instructor John Knox Laughton in a public report in 1875.)[1] The first Director of Studies was the distinguished mathematician Thomas Archer Hirst, who is the subject of Chapter 7, and the mathematical staff comprised a Professor, an Assistant Professor, two Instructors, and two Assistant Instructors.

The first Professor of Mathematics was Robert Kalley Miller (1842–1889), a Cambridge man, who had gained his BA (aegrotat) in 1867 (he had been thrown from a carriage shortly before the Tripos) and had been first Smith's Prizeman the same year. His salary at Greenwich was £600, and he seems to have been able to combine his Greenwich duties with the post of Bursar at his old Cambridge college, Peterhouse. He was the author of a popular book, *The Romance of Astronomy*, published by Macmillan in 1873, with a second edition in 1875 and a third in 1882.[2]

The Assistant Professor, whose salary was £500, was Carlton John Lambert (1844–1921), who had been Third Wrangler at Cambridge in 1867.[3] He was promoted to Professor in 1877 and served until his retirement after 36 years of service in 1909, at which point the second Professorship was discontinued, leaving William Burnside as sole professor.[4] One of Lambert's pupils was Earl Jellicoe (1859–1935), the future Admiral of the Fleet who commanded the Grand Fleet at the battle of Jutland, who recalled of his time at Greenwich

> There were nine of us in the class. The work was hard, and towards the end of the course I worked frequently up to midnight. I found time, however, during this period at the college to play a good deal of rackets and tennis, games of which I was particularly fond, and also to play rugby football as a wing forward in the College fifteen ... Our Royal Naval College fifteen was a strong side, and we played many good football teams including Richmond, Blackheath, and Oxford university in the days when Rotheram and Wade were in the Oxford team. During my course I owed a great deal to Mr. Lambert, our instructor in mathematics at Greenwich, and was pleasantly surprised when I passed first in the examination and was awarded the £80 prize, less income tax, which deduction I found difficult to understand.[5]

This had been, in 1882–1884, Jellicoe's second period at the Royal Naval College, during which he qualified as a gunnery lieutenant. He had previously in 1877 gained first-class certificates in each of navigation, gunnery and torpedo, an achievement which should have gained him immediate promotion to lieutenant, but because four others in the class had had the same success and he had been the most junior, he had had to wait eight months for the promotion.[6]

The two Instructors, at a combined salary of £800, were J. Oborn and John Knox Laughton (1830–1915).[7] Laughton (Figure 6.1), who was knighted in 1907, won subsequent renown as a naval historian: according to one historian, he was to transform the study of naval history from 'little more than "romantic fiction" to rigorous modern scholarship'.[8] He was a Cambridge Wrangler in 1852, but three other men from his college Caius came above him, which limited his prospects, and he took a position as a Naval Instructor in the Royal Navy. After some years at sea, in 1866 he became an Instructor at the Royal Naval College, then at Portsmouth, and moved to Greenwich with the College. In 1873 and 1874 he published two important papers, an *Essay on Naval Tactics* and *The Scientific Study of Naval History*.[9] From 1876 he lectured additionally at Greenwich on naval history (for which his pay was increased by £31 10s.).[10] After Easter 1885 his services as Mathematical Instructor were no longer required and he chose to retire rather than be retained on half pay. He kept his position at Greenwich as Lecturer on Naval History and continued in that role after he was subsequently appointed Professor of Modern History at King's College, London.[11]

The assistant instructors appointed in 1873 were J. Henry and Richard Wormell (1838–1914), with salaries of £300 each.[12] Wormell was a protégé of Hirst, a former mason and bricklayer who had lodged with the mother of William Barrett, assistant at the Royal Institution. Inspired by Barrett, he had derived a theorem on Foucault's pendulum. This had impressed Hirst, whose encouragement led him to take the London MA in mathematics, in which he achieved the top marks. While teaching at the Central Foundation School in Finsbury, he gained an external BSc in 1868 and in

FIGURE 6.1   John Knox Laughton. (© Royal Meteorological Society.)

the same year published the first two of his 25 textbooks. He stayed for only a year at Greenwich before returning to the school as headmaster. He was a founder member in 1871 of the Association for the Improvement of Geometrical Teaching (which was renamed the Mathematical Association in 1894), was its President for 1893 and 1894, and was editor of the *Educational Times* for ten years from 1883.[13]

6.3 The Mathematical Curriculum at the Royal Naval College

The 'Naval University' at Greenwich was not without its critics. Laughton's close friend Commander Cyprian Bridge (later Admiral Sir Cyprian) wrote to him a long letter about the curriculum in 1877:

> I am going to make a hot attack upon views as to the College *curriculum*; so be prepared for the worst. [...]
>
> How are we poor devils who have, or have had the responsibility of taking Her Majesty's Ships about the world and disciplining and training their crews got an ineradicable idea into our heads, from which all the professors

on earth will not extract it by any process of evolution known to man, that we want our officers to be seamen first, then gunners, then perhaps engineers, then perhaps surveyors, and – no one objects – Algebraists and arithmeticians about last of all? I should say a good linguist is of infinitely more use in the Navy, and would find his services more frequently required, than five senior wranglers. I want to have the Senior Wranglers too: there is room even for these poor outcasts in our Great Navy, but I do not want to have a service composed of senior wranglers and failures to become such. Yet that is what the Professors (I use the term as being the most opprobrious I can think of) desire to make of the young officers of the future. [...]

Now I will go further than even the professors (to whom this time I will not give a capital initial) and say we do want Civil Engineers, Naval Architects, even naturalists, geologists and students of languages: there is room for all, and the greater diversity the better. Had sailors been able to make Keyham and Devonport yards they would certainly not have made them without a safe landing place for a boat during the prevailing winds. But we do not want our good practical men turned out of the service because they are not Isaac Newtons or Rowan Hamiltons. The two best midshipmen I know are nearly sure to fail at the College 'till the service is quit of them, and much it will lose when they go. [...]

I know you will say that at College Mathematics is not the only subject taught. Say it, and yet you will not refute us; for its importance overshadows all others. If you have had a heart hard enough to read as far as this you will think me too great a bore. I hope I have not been abusive; but my treatment here has developed a fine capacity for swearing and I will not say that I have not occasionally used strong language; but I feel strongly, and even as I quit it, I hope to do some little good to a service in which I have wasted a life and ruined a career.[14]

Others also felt that a focus on high academic standards was leading to the loss of good practical men. 'A Naval Nobody', writing about naval education in 1878, wanted seamen to be kept 'longer studying *on shore*, at a *bona fide* college, and not at a farce as is the naval college at Greenwich. For that that college is a farce no-one who has studied there will deny'. He calls for the teaching of practical science – 'Vary the dull round of x, y, and z, of Euclid, of dry mathematics, with the knowledge of nature and her laws'. For

> You hammer only those subjects into our heads which no sooner are we free to drop them than we do so like a hot potato. What then becomes of your x, y, z's, the hunt after which has ended at last? They have run to earth; there let them stay, for it will not be we who dig them up. What becomes of your hydrostatics, which appear not to have taught us the simplest principles of the science, for when we come to apply them we cannot calculate, though all the data be given, at what angle our ship will, or will not, 'Turn turtle'? What becomes of your mode of teaching geometry – Euclid, &c., &c.?[15]

Following the appointment of a committee to enquire into the Royal Naval College in 1877 the Admiralty decreed a number of changes, reflecting a perceived need to address an excessive failure rate, especially in mathematics:

My lords consider it to be of great importance that the Examinations shall be conducted with due regard to the object in view. viz: to provide efficient Officers for the Navy and not to exclude them; and they trust that with a proper understanding of real co-operation between the Director of Studies and the Examiners, no apprehension need be entertained of either the Examiners suffering wrong, or of the Service being deprived of good Officers who should be retained in it ... While a certain amount of mathematical knowledge is indispensable in the cases of all Officers, it is very desirable that such as have not any particular aptitude for the study of the higher Branches of Mathematics, should have every facility for employing their time in the acquisition of knowledge in other subjects which are more within the scope of their talents, and which will prove valuable to them in the discharge of their professional duties.[16]

Examinations were to be conducted solely by examiners not connected with the College, registers should be taken at lectures, and Director of Studies should frequently visit classes without notice. Further Professors and Assistant Professors were to be employed in place of Naval Instructors (this change led to Laughton's retirement), but 'My lords do not consider that a system of payment by results could be applied successfully to the Teaching Staff'.

The archives contain documents showing other concerns in the Naval College at that time: preventing cheating in the examinations, dealing with a student who had deliberately failed and another who had fallen asleep during his examination, and the misconduct of two Acting Sub-Lieutenants 'in introducing women of immoral character into the Royal Naval College'.[17] They also testify to the increasing difficulty in filling the double office of Chaplain and Naval Instructor because, with the availability of other degrees at Cambridge, fewer men intending to take Holy Orders were passing the Mathematical Tripos.[18]

6.4 Subsequent Mathematics Professors and Instructors

There was a rapid turnover in the mathematical instructor positions. During 1879 one of the then Instructors, the South African Richard Prince Solomon (1850–1913), another Peterhouse man who had been twenty-third Wrangler in 1871, took over Miller's duties when the latter was granted two months' sick leave. Solomon, who was knighted in 1901, was to become a prominent politician in his native land, where he 'exercised his powers of persuasion, his moderating influence, and his industry in the work of reconstruction' after the South African War.[19]

On Hirst's retirement in 1883 he was replaced as Director of Studies by William D. Niven (1843–1917), who had been third Wrangler in 1866. Niven, who worked on spherical and ellipsoidal harmonics, was a friend of James Clerk Maxwell and edited Maxwell's scientific papers, published in 1890: he was President of the London Mathematical Society from 1908 to 1910. When Miller retired in 1885, Niven brought William Burnside (Figure 6.2) to Greenwich as Professor of Mathematics. Burnside is the subject of Chapter 8.

Another prominent mathematician at Greenwich in the early twentieth century was Sir James Blacklock Henderson (1871–1950), who was Professor of Applied Mechanics

FIGURE 6.2   William Burnside. (Courtesy of the London Mathematical Society.)

from 1905 to 1920. Henderson had studied at Glasgow and Berlin Universities. He was a prolific inventor who during his time at Greenwich was granted patents for work on projectiles, propellers, bomb sights, and gyroscopic controls for gun mountings.[20] During the War, Henderson was one of the mathematicians employed in Room 40, the Admiralty's cryptographic operation.[21]

When Burnside eventually retired in 1919, his term having been extended because of the war, his successor was originally going to be James Mercer (1883–1932), who was working at the College as an Instructor and who had seen wartime action at Jutland. However the salary attached to the professorship, initially £500, was uncompetitive: those advertised for Chairs at Birmingham and Cambridge were £1,000 and £1,350. It was proposed in March to increase the Greenwich salary to £800, but while the negotiations with the Treasury were proceeding, Mercer, having no definite answer by July, arranged instead to return to Cambridge.[22] Mercer, who had been bracketed Senior Wrangler with J.E. Littlewood in 1905 and was Smith's Prizeman in 1908, was described by the great English mathematician G.H. Hardy as 'a mathematician of high originality and great skill' who, however, suffered poor health and after the war never fully realised his potential.

The next Professor of Mathematics was instead to be Charles Godfrey (1873–1924) (Figure 6.3). He was Fourth Wrangler in 1895, after Bromwich, Grace, and Whittaker (all subsequently distinguished mathematicians), and was later awarded the Isaac Newton Scholarship.[23] After a period as Senior Mathematical Master at Winchester College, where 'he completely revolutionised the old system',[24] he was appointed

CHARLES GODFREY, M.V.O., M.A.

FIGURE 6.3 Charles Godfrey. (Courtesy of the Mathematical Association.)

Headmaster of the Royal Naval College, Osborne, a junior officer training college for the Royal Navy for boys from the age of 13.

In 1901 Godfrey, with his former student Arthur Siddons, had written a letter on mathematics education reform which, signed by twenty others, was to become known (either ironically or with a slightly surprising lack of concern for arithmetic) as 'The Letter of the 23 Schoolmasters' when it appeared in *Nature* in January 1902. This led to the formation of the Mathematical Association Teaching Committee. Godfrey was to serve on this committee from its formation until his death:

> His influence was stronger than that of any other member; in some cases he
> was the power and stimulus behind the active work done by other members
> of the Committee, on other occasions he did the active work himself.[25]

In 1871 the Association for the Improvement of Geometrical Teaching had been founded, with Thomas Archer Hirst as its first President and Rawdon Levett, later Godfrey's teacher at King Edward's School, Birmingham, as Secretary. (The AIGT became the 'Mathematical Association' in 1897.) Although by 1888 Oxford and Cambridge allowed in their examinations proofs other than Euclid's, the order of Euclid's propositions was still sacrosanct. Arthur Cayley, one of the leading mathematicians of his day, who argued 'Euclid must stay!' had enormous influence: a member of the Board considering the matter reported 'We cannot go against Cayley'.[26] But Cayley's successor as Sadlerian Professor at Cambridge, A.R. Forsyth, favoured examination reform. Forsyth encouraged Godfrey and Siddons to write an elementary geometry textbook for schools and himself astutely piloted the required changes through the Cambridge University Senate. The textbook was published in 1903 and sold 22,000 copies within ten months.[27] Godfrey wrote later of the consequence:

> The effect seemed to us at the time to be very great; but I suppose that as a
> matter of fact a great majority of schools went on teaching in the old way,
> and no doubt many to this day have been quite unaffected by the new spirit.
> But in those schools which moved with the times the inevitable effect was
> probably rather chaotic. This was natural and quite inevitable.[28]

Godfrey and Siddons subsequently collaborated on *Elementary Algebra*, published in 1912 and 1913, and then on a book of *Four-Figure Tables* (1913), which had achieved sales of almost 4 million by 1973.[29] During the War Godfrey, like Henderson, worked in Room 40: his younger brother John Henry Godfrey (1888–1971) was to be Director of Naval Intelligence during the first part of the Second World War.[30]

Godfrey was appointed in August 1920 and took up the Chair in May 1921, at a salary of £600. He took enthusiastically to the teaching of advanced work at Greenwich: but he died suddenly in 1924 from bronchial pneumonia. Siddons wrote that 'He was not a man who pushed himself forward, and it is only those who knew him intimately who can appreciate how much of the reform of mathematical teaching in the last twenty years is really due to his influence'.[31]

In the gap between the departures of Burnside and Mercer and Godfrey's taking office George Arthur Witherington (1873–1942), the Instructor in Mathematics, had been Acting Head of Department, and in June 1921 he was promoted to Assistant Professor.[32]

After Godfrey's death Witherington replaced him as Professor. Witherington had gone to Queens' College, Cambridge, in 1899 (at the age of 26), was awarded a scholarship in 1901, and graduated with First-Class Honours in Mechanical Engineering in 1902. He doesn't appear to have been, in today's terms, research-active in mathematics.[33]

Louis Melville Milne-Thomson (1891–1974), a Wrangler in 1913 (by which time the Cambridge Wranglers were no longer ranked), joined the Royal Naval College as Instructor in mathematics at a salary of £500:[34] he was to be Professor of Mathematics from 1934 until his retirement in 1956. In 1923, in addition to his mathematical duties, he was giving lectures on Spanish (for which he received extra payment) and Dutch (unpaid).[35] He worked on mathematical tables, publishing *Standard Four-Figure Tables* in 1931 with Leslie Comrie (of whom more in Chapter 9), and *Jacobian Elliptic Function Tables* in German in 1931 and in English in 1950; he also wrote textbooks on *The Calculus of Finite Differences, Theoretical Hydrodynamics*, and *Theoretical Aerodynamics*. From 1946 he was Professor of Geometry at Gresham College in the City of London, an institution in the City of London founded under the will of Sir Thomas Gresham in 1597, whose professors present regular popular lectures to the general public.[36] The Gresham Chair of Geometry is the oldest mathematical Chair in the UK.

Milne-Thomson's elegant Circle Theorem in fluid dynamics allows one to write down, without calculation, the change to a two-dimensional flow when a circular obstruction is introduced: it has been praised for its elegance by one of today's fluid dynamicists who described it as 'a theorem that belongs to an earlier age'.[37]

Milne-Thomson's reputation was such that Hugh Michael, later to be a distinguished mathematician at University College, London, wrote of his disappointment, when he came to the Royal Naval College as a Royal Navy Instructor Lieutenant, that he did not at that time meet Milne-Thomson because, 'For some reason which I did not learn, it seemed that Academic staff of the college did not have contact with service personnel on short courses of instruction like mine'.[38] (Michael did meet Milne-Thomson much later, at applied mathematics conferences in Australia, after Milne-Thomson had retired from Greenwich.)

In 1952 Milne-Thomson brought the fourth British Mathematical Colloquium to the Royal Naval College. The attendance was well over 100.[39] The meeting was divided into three sessions. Tuesday 29 July was devoted to Analysis, with J. C. Burkill (1900–1993), E. R. Reifenberg (1928–1964), and F. F. Bonsall (1920–2011) talking in the morning and W.K. Hayman (born 1926) speaking on 'The growth of integral functions' in the evening. The next day covered Geometry, with talks by D. G. Northcott (1916–2005) and the invited speaker, the distinguished algebraic topologist Heinz Hopf (1894–1971) from Zurich, who spoke on 'Metric, analytic and algebraic structures on manifolds'. On Thursday 31 July the morning session on Algebra featured B. H. Neumann (1909–2002), K. A. Hirsch (1906–1986) and D. R. Taunt (1917–2004), with H. Davenport (1907–1969) giving the evening lecture on 'Diophantine approximation'. Davenport's talk was chaired by the distinguished mathematician Louis Mordell (1888–1972), and Mordell fell asleep during the talk. Awakened by the applause, Mordell had the misfortune to tell an anecdote about Newton which he did not realise Davenport had just related.[40]

The minutes of the BMC general meeting record that the following motion was passed unanimously:

> The British Mathematical Colloquium wishes to express the gratitude of its members for permission to hold the fourth colloquium at the Royal Naval College and their appreciation of the kindness and hospitality with which they have been treated. The Royal Naval College has provided excellent facilities for the mathematical business of the Colloquium. Moreover the privilege of meeting among such surroundings as these is a notable contribution to the welfare of British mathematics.[41]

One consequence of the hosting of BMC was that the College had to provide accommodation for the two female members of the Colloquium. One of them, Hanna Neumann (1914–1971), recalled that this was handled by the two guests being given temporary naval rank for the duration of the meeting, as a means of circumventing Naval regulations.[42] The College staff were, apparently, used to military discipline: Douglas Northcott used to tell the story that the porters were shocked when he arrived late for one of the talks, such behaviour not being expected at the College.[43]

Milne-Thomson, who 'had a rather impressive presence which was made more so by his wearing a monocle',[44] retired in 1956, after which he travelled widely, including ten years as Professor of Applied Mathematics at the University of Arizona. He was succeeded as Professor at the Royal Naval College by Thomas Arthur Alan Broadbent (1903–1973) (Figure 6.4), who had joined as Assistant Professor in 1935 and retired in 1968.

A Cambridge Wrangler, Broadbent had started research under Littlewood, but he was more interested in teaching. He had wide mathematical interests and was Editor of *The Mathematical Gazette*, the journal of the Mathematical Association, from 1931 to 1955, and was the Association's President for 1953–1954. His wife Janet was the younger sister of the mathematician W.V.D. Hodge (1903–1975), who succeeded Broadbent as President of the MA: Broadbent's cat Hodge however was named after Doctor Johnson's cat rather than his brother-in-law. As well as succeeding Milne-Thomson in the Greenwich Chair, Broadbent also succeeded him as Gresham Professor of Geometry, which Chair he held until 1969. According to the subsequent Gresham Professor, and his next-door neighbour in Greenwich for thirty years, Clive Kilmister (1924–2010), Broadbent was 'the kindest, most considerate man I have ever known'.[45]

A. G. (Geoffrey) Howson joined the College in 1957 (he was subsequently a member of the School Mathematics Project, which had a profound influence on the teaching of mathematics in British schools from the 1960s, and was President of the Mathematical Association in 1988). He recalls that in his time there was a mathematics staff of six civilians and one naval officer, with Basil Brown as Assistant Professor to Broadbent. Mathematics was taught to trainee naval constructors, to midshipmen studying for an Engineering degree as external University of London students, and to nuclear engineers, whom Howson recalls as the most gifted students. There were short courses on meteorology and gunnery. Operations research was taught as an approach to the problem of a ship locating an enemy submarine. 'This was based

Photo : Dalby

THOMAS ARTHUR ALAN BROADBENT, M.A.
President, April, 1953—January, 1954

FIGURE 6.4   T. A. A. Broadbent. (Courtesy of the Mathematical Association.)

on the, at that time reasonable, assumption that the surface ship would be faster in the water than the submarine. The arrival in Portland Bay of *Nautilus*, the first US nuclear powered submarine, meant that new notes had to be written!'[46]

The atmosphere was relaxed. On Derby Day the College had a holiday with everyone repairing to Epsom.[47] Howson was a member of a lunchtime musical trio. He has fond memories of the College: 'Teaching at RNC at that time was especially rewarding because one was teaching people (all male at that time) who were about one's own age and with whom one mixed, talked, played, ate and drank after the day's lectures were over'.[48]

6.5 The Departure of the Royal Naval College

By the 1960s the needs of the Navy had changed, and its inability to award academic degrees was deterring potential students. An application was made to the Council for National Academic Awards for the power to award degrees, but this was rejected, and the Howard-English Committee on Naval Education recommended the closure of the College.[49] The then Secretary of State for Defence, Denis Healey, decided in 1967 that the various naval training establishments should merge at Shrivenham, but discussions were prolonged and it was only in 1998 that the Royal Navy finally moved out of what is now the Old Royal Naval College.[50]

6.6 Greenwich and the Popularisation of Mathematics

It is not entirely surprising that so many of the people employed by the Royal Observatory and by the Royal Naval College at the end of the nineteenth and in the early twentieth centuries had similar backgrounds. The English education system for much of the period in question was such that a budding astronomer was likely to take the mathematical tripos at Cambridge, and so one would expect that Astronomers Royal and Chief Assistants would generally be Wranglers and Smith's Prizemen, and likely Isaac Newton Students too. In fact the first twentieth-century Astronomer Royal since the introduction of the Tripos who was not a Cambridge Wrangler was Sir Martin Ryle (Astronomer Royal 1972–1982), who had read Physics at Oxford.

One theme that has emerged from the preceding discussion of the mathematicians at Greenwich, both at the Observatory and at the Royal Naval College, is the explicit interest many of them showed both in the public dissemination of mathematics and astronomy, and in mathematical education in general. It was perhaps a consequence of the 'peculiarly British phenomenon'[51] of the position of Astronomer Royal that there was public interest in both the holder of the position and the subject, and, as we have seen from the Times's response to the observations of the eclipse of 1919, there was public excitement over contemporary developments.

Nevertheless it is notable that the many mathematicians at the Royal Naval College were also active in these aspects of mathematical life. While Burnside's *Theory of groups of finite order* was a monograph rather than a popular book, he no doubt hoped that it would attract students to his subject, and in his Presidential Address to the London Mathematical Society he again attempted to stimulate interest in group theory. Miller's popular book on astronomy went through three editions in the UK, and Milne-Thomson's books were important and widely read guides to the developing fields in mathematics. That both Milne-Thomson and Broadbent took on the post of Gresham Professor of Geometry – a position whose raison d'être was to give frequent public lectures – shows their commitment to promoting the public understanding of mathematics.

It is also striking that so many of those mentioned above took a very active interest in mathematics education. From its foundation to the departure of the Royal Naval College Greenwich mathematicians were closely involved with the Association for the Improvement of Geometrical Teaching and its successor, the Mathematical Association. Hirst and Wormell were Presidents of the former, and Turner, Eddington, Chapman, Spencer Jones, Broadbent, and Howson were all Presidents of the latter.

Greenwich mathematicians regularly participated in the International Congress of Mathematicians: Burnside at Cambridge in 1912, B.P. Haigh (Professor of Applied Mechanics at the Royal Naval College), Henderson, Jackson, and A. Beale (Whitworth Senior Scholar at the Royal Naval College) at Toronto in 1924,

Haigh and Milne-Thomson at Bologna in 1928, and Milne-Thomson at Zurich in 1932.[52] Howson recalls that Broadbent took all his staff to the ICM at Edinburgh in 1958 and actively encouraged them to participate in the Mathematical Association and London Mathematical Society.[53] Burnside, Niven, and Chapman all served as Presidents of the London Mathematical Society. It is clear that staff at both Greenwich institutions were active participants in national and international mathematical circles.

6.7 Conclusion

This chapter has presented some of the remarkable characters who taught mathematics at the Royal Naval College in Greenwich. Some made their careers as mathematicians, while others went on to gain distinction in other fields. Some, like Burnside and Milne-Thomson, made major contributions to twentieth-century research mathematics; others, like Godfrey and Howson, contributed to important developments in mathematics education; several were active in public engagement; and many gave significant service to the mathematics community. During its time at Greenwich the Royal Naval College, through the activities of its staff, played a full part in mathematical life in the UK.

7

Thomas Archer Hirst at Greenwich, 1873–1883

Robin Wilson and J. Helen Gardner

CONTENTS

7.1 Introduction

On 29 November 1872 Thomas Archer Hirst F.R.S.,[1] Assistant Registrar of the University of London (Figure 7.1), sought advice from William Spottiswoode,[2] printer to the Queen and future President of the Royal Society. The topic of their conversation was a letter that Hirst had just received from his friend Thomas Huxley,[3] the well-known supporter of Charles Darwin.

> My dear Hirst,
>
> I dare say you may have heard that the Admiralty are about to establish a sort of Naval University at Greenwich, for the purpose of giving the Engineers and persons employed in the Dockyards, as well as the pick of the Executive officers, instruction in those branches of mathematical and physical science with which they need to be acquainted.
>
> As the Students will be mainly Naval officers, the President of the Institution will be a Naval officer, and Admiral Cooper Key is to be appointed to that office.
>
> But the Admiralty intend to appoint a resident Principal to whom they would look for the organization and supervision of the studies of the place. They are ready to give this Principal, or Director of studies, £1000 a year, a house, and a retiring pension – and they are very much disposed to be guided by his opinion as to the extent to which he should himself take part in the teaching.

FIGURE 7.1   Thomas Archer Hirst, first Director of Studies at the Royal Naval College. (Courtesy of the London Mathematical Society.)

I apprehend they would be content with his delivering a course of lectures on a subject selected by himself and would not expect him to perform tutorial or examining duty.

They want a man of scientific eminence to give the institution dignity with the public; a man who can understand and attract young men and get them to enter with heart into their work, and who is a good judge of their capacity and fitness for future employment; and, finally, a man who has sense, tact and knowledge of the world sufficient to ensure his keeping things straight between the body of professors and the executive head of the Institution.

Now, in my opinion and that of another discreet friend of yours [Spottiswoode], you are just the man the Admiralty want: and I write to ask if I may tell the First Lord that you are disposed to entertain a proposal to fill the office.

Let me beg you to think over the matter carefully and consult Spottiswoode before you decide.

Ever yours very truly,

T. H. Huxley.

The 'sort of Naval University' to which Huxley referred became the Royal Naval College at Greenwich,[4] housed in the magnificent buildings designed by Christopher Wren in 1694 as a home for sick and aged mariners. It had continued as the Royal Hospital for Seamen, reaching its peak around 1813 when it housed 3,000 pensioners. But with long years of peace, the hospital declined and eventually closed in 1869. Meanwhile, a new Naval College had been established in Portsmouth in 1841 as scientific instruction was proving to be highly inadequate at sea in an age of transition from sea to steam. With the availability of the Greenwich site, the Admiralty proposed to re-locate the Royal Naval College in order to form a university for the Royal Navy. The new College was duly established by Order in Council on 16 January 1873:

We do therefore beg leave to recommend that Your Majesty will be graciously pleased, by Your Order in Council, to approve of the closing of the Royal Naval College, Portsmouth, and the founding of a College at Greenwich, to be styled "The Royal Naval College," accordingly; and the establishment of such regulations as regards the admission of Students and courses of studies, as may appear to us, from time to time, to be desirable.

On 30 January, the following order was issued by the Admiralty[5]:

The College, subject to the subjoined regulations, will be open to officers of the following ranks:–

1. Captains and Commanders.
2. Lieutenants.
3. Navigating Officers.
4. Naval Instructors.
5. Acting Sub-Lieutenants and Acting Navigating Sub-Lieutenants.
6. Officers Royal Marine Artillery. Ditto, Royal Marine Light Infantry.
7. Officers of the Engineer branch, viz: Chief Engineers, Engineers, 1st Class Assistant Engineers.
8. A limited number of Dockyard Apprentices will be annually selected by competitive examination, for admission to the College.

A course of instruction at the College will also be open to a limited number of

9. Private Students of Naval Architecture or Marine Engineering.
10. Officers of the Mercantile Marine.

It is not intended to provide at Greenwich for the education of the Naval Cadets.

The purpose of the College was 'to provide the most efficient means of higher education of Naval officers adequate to the constantly increasing requirements of the Service', though 'without prejudicing the all-important practical training in active duties'. The training was to be largely scientific in nature, providing 'the highest possible instruction in all branches of theoretical and scientific study bearing upon the profession'.

The Governor of the College was to be the First Lord of the Admiralty. The Rt. Hon. G. J. Goschen held this position from 1873 to 1874, to be succeeded in turn by G. W. Hunt (1874–1877), W. H. Smith (1877–1880), and Lord Thomas Baring (1880–1885). The College would be run on a day-to-day basis by its President, a Flag Officer responsible to the Admiralty; the first of these was Vice-Admiral Sir Astley Cooper Key KCB FRS (1873–1876), followed by Admirals Edward Fanshawe (1876–1878), Sir Charles Shadwell (1878–1881), Sir Geoffrey Phipps Hornby (1881–1882), and William Luard (1882–1885). Assisting the President would be a Captain of the Royal Navy, responsible for non-academic issues, while the academic programme was to be the responsibility of a Director of Studies, heading an Academic Board of five professors. In addition, there would be about two dozen lecturers – both naval and civilian.

The first Director of Studies was indeed Thomas Archer Hirst. But who was he, and why did Huxley consider him so ideally suited for this important new post?

7.2 Thomas Hirst's Career, 1830–1872

Thomas Archer Hirst was born in 1830 in Yorkshire. His parents were from successful wool-merchant families and his father, a wool-stapler, was wealthy enough to retire at the age of 31. When Tom was 11, he was sent to school in Wakefield, where mathematics was his favourite subject. But his father's death from a drinking accident in 1844 meant that Tom, then aged 15, needed to leave school and find employment. He was articled to a Halifax engineer, Richard Carter, who had been commissioned to survey for the West Yorkshire railway, then under construction.

On starting his appointment, he commenced writing a diary.[6] These diaries eventually covered 45 years and chronicle the scientific circles in which he moved, the people he met, and, in particular, his experiences at the Royal Naval College.

Carter's principal surveyor was John Tyndall,[7] who would later play an important role as Michael Faraday's successor at the Royal Institution. Tyndall and Hirst developed a close friendship that would have a great influence on Hirst's life. In 1848, Tyndall left Halifax to study at the University of Marburg in Germany. Hirst went out to visit him, but the visit had suddenly to be cut short by the unexpected death of his mother. He was now financially independent and resolved that railway surveying should not be his chosen career.

Attracted by Tyndall's experience in Germany, Hirst returned to Marburg to work towards a doctorate.[8] His timetable was intensive, and his diaries record his motivation and dedication in studying for up to 16 hours a day. In July 1852, he was awarded his degree for a dissertation 'On conjugate diameters of the ellipsoid'.

With its completion, Hirst began a formative stage in his life that would fit him for his future role at Greenwich. British mathematics lagged far behind that of the

Continent, and Hirst decided to travelled widely in Europe, calling on mathematicians wherever he went – first to Göttingen where he visited Gauss and conducted magnetic experiments with Weber, and then to Berlin where he attended the lectures of Dirichlet and Steiner.[9]

Hirst returned to England in mid-summer 1853. Although there were limited job opportunities for a young mathematician, he was offered a teaching post at Queenwood College, near Salisbury, where Tyndall had taught before his Marburg days. Queenwood encouraged practical skills, and Hirst's geometry teaching involved surveying rather than the usual learning of Euclid's *Elements*. This approach would be reflected later when he emerged as a reformer of the teaching of geometry in schools.

While in Marburg, he had met a young lady called Anna Martin. They were married in 1854 and returned to Queenwood. But Anna soon began to show signs of advancing tuberculosis. Hirst resigned his teaching post to devote himself to her, and they travelled in the South of France, vainly searching for a cure. In July 1857, Anna died, and he never completely recovered from the tragedy.

Hirst decided to devote his time entirely to his geometrical researches. His diaries record lengthy periods in Paris where he became acquainted with Liouville, Chasles, and Bertrand, and in Italy where he encountered Brioschi and Cremona.[10]

After two years establishing his reputation on the Continent, Hirst returned to London where he attended lectures of importance, such as those by Huxley on Darwin's recent theory of evolution, and met the major figures of English mathematics, such as Cayley, Sylvester, and Boole.[11] Employment for a mathematician was still difficult to obtain, but again Hirst fell on his feet, becoming a teacher at University College School in Gower Street. Shortly after, in June 1861, at the age of only 30, Hirst was elected to a Fellowship of The Royal Society, his certificate being signed by (among others) Boole and Sylvester.

In 1863, deteriorating health made work impossible, and he set off once more to travel in Europe. His diaries record meetings with old and new acquaintances, including the Russian mathematician Tchebichef[12] who was later to visit Greenwich.

By November 1864, he was home, and Hirst, Tyndall, Huxley, Spottiswoode, and other close friends formed themselves into a scientific club of nine members. United by their devotion to science, pure and free and untrammelled by religious dogmas, the 'X-club' was to influence the organization and image of English science for the next twenty years.[13]

Hirst's career was developing rapidly. In late 1864, he was elected to the Council of the Royal Society. Shortly after this, the first meetings took place of what would become the London Mathematical Society, with Augustus De Morgan[14] as Founding President. Hirst was elected a Vice-President and was a member of its Council for almost two decades, becoming Treasurer and then President. By this time, he had given up his teaching to devote more time to geometry. But in August 1865, he was appointed Professor of Mathematical Physics at University College, establishing him as one of only seven physics professors in the country. Two years later, when De Morgan resigned his mathematics chair after a dispute with the College, Hirst took his place, becoming Professor of Pure and Applied Mathematics.

Increasingly, his administrative and lecturing duties left him too little time for his geometrical investigations. His diary entries complain that others were now working in areas to which he had previously contributed while he had time only to look on. He longed for leisure to pursue his researches.

Eventually, he resigned his chair and applied for a well-paid administrative job that he hoped would give him more time. In March 1870, he became Assistant Registrar of the University of London, hoping to 'come in contact with good and influential men and thus be able to influence to some extent the character of education in England'. Increasingly, he was involved with educational matters and attended meetings on education in secondary schools. His aim was to see Euclid's *Elements* replaced as a school text-book, and his years as a surveyor, together with his experience of teaching practical geometry at Queenwood College and University College School, left him well placed to help establish the Association for the Improvement of Geometrical Teaching.[15] Hirst was its first president and held office for seven years.

By 1873, Thomas Hirst had become a major figure in Victorian science. As President of the London Mathematical Society, a Vice-President of the Royal Society, an ex-professor of University College (twice over), and a member of the famous X-club of radical scientists, he was a personage of great influence and stature. With his long experience of teaching at both school and university level, Hirst was the ideal candidate to 'understand and attract young men', and to be 'a good judge of their capacity and fitness for future employment', as Huxley had believed.

7.3 Hirst's Appointment at Greenwich

In his November 1872 meeting with Spottiswoode, Hirst had confided that[16]:

> The appointment is a high one. All I fear is the absorption and worry – both ruinous to my Geometry which is dearer to me than all, even than the Admiralty's offer of £1000 a year and a house.

After spending a wakeful night considering the Admiralty's offer, he sought an interview with the First Lord, the Rt. Hon. George Joachim Goschen,[17] to discuss the appointment:

> I called accordingly, was shown into his dressing room where I found him in his dressing gown on the sofa. He explained to me the nature of the appointment, and I told him that the only hesitation I had in undertaking its duties arose from my fear that they would be so absorbing as to prevent me from pursuing my own scientific work. He said he should himself greatly regret this but scarcely feared it.

Hirst was also concerned about whether the position of Director of Studies would require him to give lectures:

> I dwelt long on the question of being expected to lecture, and I expressed the hope that this would not be made a condition of acceptance. He replied that he could not commit himself to a promise that no lectures would be required from me, but he admitted that they would be guided in a great measure by my own opinions after the College had been fairly started.

Hirst visited Admiral Key[18] at the Admiralty and was impressed with him and his ideas for the College. Two days later, they went to Greenwich to inspect the College buildings and residences, and a few days later, Hirst returned to Goschen to confirm that:

> I had no longer any hesitation about acceptance, that the more I had thought of the work the more I had warmed towards it and that my ambition would be to make the College second to none in existence.

However, a few matters still needed attention, to all of which Goschen responded positively:

1. That the work of organization and supervision would occupy me fully without lecturing. (He asked me not to make this conditional but an expression of my strong opinion)
2. That the end of February would be the most suitable time for my leaving the University and that between then and October I should be glad of visiting the existing Naval departments in England and the Continent so as to familiarise myself by actual inspection with the duties which the several classes of students would on leaving the College be called upon to perform. (This proposition met with his full concurrence)
3. That the house I had seen was scarcely roomy and airy enough for a permanent residence. (He had already heard this from Admiral Key and he assured me that as far as he could everything should be done to meet my wishes in this matter)
4. That it was usual in estimating the retiring pension of persons in such positions as this to add a certain number of years to those of actual service. (He replied promptly by saying that he had already made a note of this as a subject to be mentioned to the Chancellor of the Exchequer).

On 18 December, Hirst duly announced to the University of London Senate that he would resign the Assistant Registrarship at the end of February. On 6 January 1873, Goschen formally offered him the post, but with the annual salary increased from £1,000 to £1,200 'due to the high position we wish you to occupy and more suitable when compared to the other salaries proposed'. Hirst would thus be receiving twice what he had earned at London University, with a house included as well. He sent in his acceptance on 12 January 1873, confirming that he would commence his new duties on 1 March and continuing:

> Permit me to assure you, Sir, that I appreciate highly the distinctions you propose to confer upon me, as well as the very kind and courteous manner in which your offer has been conveyed to me. Let me add too, that I have no higher ambition than to be able to carry out, thoroughly and successfully, the well matured plans you have formed for providing officers of the British Navy with the very best Scientific and Technical Training now attainable.

The appointment was announced in *The Times* on 23 January and was well received. The following verse was written 'By a Scientific Person; on seeing a certain announcement in to-day's paper'.[19]

Three Cheers for T. G.!
Who has shewn he can see
What we saw from the first,
    The merits of Hirst.
              May his vessels all speed,
              And his Navy succeed,
Till England a splendid example affords;
May his College at Greenwich
Prove worth all the tin which
Is so gen'rously spent on it now by
              My Lords!

7.4 The College Opens

The College was to have five professors, in Mathematics, Applied Mechanics, Physical Science, Chemistry, and Fortification. On 1 January 1873, Hirst had an interview at the Admiralty with Mr Goschen (Figure 7.2), Admiral Key (Figure 7.3), and Dr Woolley (the Admiralty's Director of Education) to discuss the professorial appointments[20]:

> In Physics the qualifications of Carey Foster, and W. C. Adams were first considered, then I mentioned Reinold, Fellow of Merton College Oxford as one of whom I had heard very favourable accounts. It was determined that I should make further enquiries about him. I strongly recommended the appointment of Debus to the Chair of Chemistry. The recommendation was favourably received.

But a difficulty was beginning to emerge as Admiral Key expressed his concerns about giving too much of a German character to the College:

> He observed that I had taken my degree in Germany. Reinold was obviously of German extraction, and Debus a decided German, to which Mr Goschen added that he himself had a like origin. On this account I stated that I would not name the best man I knew for the Chair of Mathematics since he was also a German (Henrici). It was agreed that we should try to find a Cambridge man for the Mathematical Chair.

William Reinold duly became Professor of Physics, and Heinrich Debus, an old friend of Hirst's from Marburg and Queenwood College days, was appointed Professor of Chemistry. Kalley Miller of St. Peter's College (Peterhouse), Cambridge, was later appointed Professor of Pure Mathematics.[21]

For the next few months, Hirst was very busy designing syllabuses and sorting out arrangements, while commencing a regular round of visits to inspect the Dockyard Schools at Chatham, Portsmouth, Pembroke, Sheerness, and Devonport, and examine the naval cadets at Dartmouth. The list of subjects to be taught at Greenwich gradually emerged[22]:

PUNCH'S FANCY PORTRAITS.—No. 44.

RIGHT HON. G. JOACHIM GOSCHEN, M.P.

THIS IS A JOKE-'IM GOSCHEN PICTURE OF A WISE MAN FROM THE EAST, AT PRESENT ASCERTAINING WHICH WAY THE WIND BLOWS.

FIGURE 7.2   A Punch caricature of the Rt. Hon. George Goschen as a parrot on a perch whose allegiances shift according to circumstances.

Subjects

1. Pure Mathematics, including Co-Ordinate and Higher Pure Geometry, Differential and Integral Calculus, Finite Differences, and the Calculus of Variations.

FIGURE 7.3   Admiral Key.

2. Applied Mathematics, viz., Kinematics, Mechanics, Optics, and
   the Theories of Sound, Light, Heat, Electricity, and Magnetism.
3. Applied Mechanics, including the Theory of Structures, the principles of Mechanism, and the Theory of Machines.
4. Nautical Astronomy, Surveying, Hydrography, with Maritime
   Geography, Meteorology, and Chart Drawing.
5. Experimental Sciences:–
   a.   Physics, viz., Sound, Heat, Light, Electricity, and Magnetism.
   b.   Chemistry.
   c.   Metallurgy.

6. Marine Engineering in all its branches.
7. Naval Architecture in all its branches.
8. Fortification, Military Drawing, and Naval Artillery.
9. International and Maritime Law; Law of Evidence and Naval Courts Martial.
10. Naval History and Tactics, including Naval Signals and Steam Evolutions.
11. Modern Languages.
12. Drawing.
13. Hygiene – Naval and Climatic.

Meanwhile, the difficulties of finding a suitable residence proved greater than Hirst had expected, the first being that 'Capt. Baker an old Naval officer (80 years of age) occupied a flat of the house upon which I had set my heart and is not to be disturbed during his life'. In early January, Hirst looked at the 'Steward's House', then occupied by the Chaplain:

> It is isolated from all other buildings in the most westerly corner of the rectangular plot of ground on which the Hospital and its grounds are situated. It is a compact house, a little too much so in fact, and has the great advantage of having a little private garden around it. By building to it a wing it might be made a very comfortable residence. I brought away the plans from Mr Loughborough the clerk of the works with a view of seeing if I could not design a suitable wing.

In the event, he was indeed assigned the Chaplain's house and moved in to his new home on 20 March, with his niece Lilly as housekeeper and hostess for his entertaining of the great and the good.

The first great social event to take place occurred in June when there was a grand reception to welcome the Shah of Persia[23]:

> Mr Goschen entertained the School at Luncheon in the Painted Hall. He, the Prince of Wales, the Princess, the Caesarevich, the Caesarovna, the Duke of Edinburgh and a large party arrived by steamer shortly before two o'clock. We residents abstained from inviting any one, Mr Goschen's guests being so numerous. Nevertheless our house was pretty well filled and Lilly entertained them very nicely. … I was at the gate-way with Mr. Goschen and Admiral Key to receive him. … The grounds were very gay indeed and the whole reception was a perfect success.

By August, Hirst could report that he had 'succeeded in organizing the Royal Naval College and setting 200 Officers to their appointed work'. A photograph, taken in the summer of 1873, shows him with his 200 students (Figure 7.4). All had been managed on time: the College opened on 1 October, according to schedule, as Hirst duly recorded:

> *The Royal Naval College Greenwich opened with upwards of 200 students.*
> There was very little confusion, and preparations for work advanced rapidly.

FIGURE 7.4   Hirst and the students at the Royal Naval College, summer 1873.

7.5 Successes and Failures

Hirst's diaries recall that 'each day from 9 A.M. to 5 P.M. I was as busy as I could be, directing, superintending and teaching'. It was indeed an exacting and time-consuming job as he organized his academic staff of 28, scrutinized the examination arrangements, visited the Dockyard Schools, and dealt with the various demands of the Admiralty; all these would become an increasing burden during his ten years with the College. In the evenings, he frequently travelled into town for gatherings of the X-club or the regular meetings of the London Mathematical Society, or for dinner with his scientific friends at the Athenaeum Club.

From his first arrival at Greenwich, Hirst enjoyed hosting distinguished mathematicians from abroad. In the few days after the College opened, he recorded[24]:

> Oct. 2: Prof. Klein of Erlangen called upon me. I invited him to dine with me at the X-club first meeting of the Session.
> Oct. 3: Tchebichef, who called on me a few days ago, and Klein dined with me at Greenwich. Tchebichef told us of a mode of converting circular into rectilineal motion (à propos of the parallelogram of Watt) which was a simple and beautiful application of Quadric Inversion.
> Oct. 8: Klein came to luncheon. We visited with Admiral Key Airy's model of the Transit of Venus.

FIGURE 7.5   The Officers' Mess of the Royal Naval College in 1881.

There were also guest lectures by distinguished visitors. In April 1874, Sir George Biddell Airy,[25] the Astronomer Royal, presented a course of three lectures on *The Magnetism of Ships*, and two years later, John Tyndall lectured on *Fog Signals* to a crowded room.

The Royal Naval College was proving to be a success. On 1 October 1874, as the College began its second year, Hirst noted that the 'College opened with about 240 students, almost as many as could possibly be accommodated'.

But it was not all work and no play, as there were social occasions at which Hirst was able to relax, such as the farewell dinner to the Captain in March 1875 when he dined at the Officers' Mess (Figure 7.5):

> The Sub-Lieutenants invited me to their common room after dinner. Prince Louis received me, and they made me dance the Lancers with a pale Sub-Lieutenant for partner! Prince Louis played the piano.

However, the job was undoubtedly a highly stressful one, and at times, Hirst became disillusioned with it, as he recorded at the end of 1875:

> A whole term's work has been got through, the Britannia visit and examinations are over, the inspections of the Dockyard Schools at Devonport and Pembroke have been made just as last year and here I am alone and this time lonely at the Gatehouse Hotel, Tenby. I am lonely simply because I am kept by official routine not from occupying my thoughts with my geometry. ... But the fact that Admiral Key is about to leave Greenwich and be succeeded in the Presidentship of the College by Admiral Fanshawe has caused me to put everything aside and push forward that long postponed programme of the studies of the Royal Naval College. I dislike the work thoroughly and

entertain but a poor opinion of its utility. Hence my loneliness. But no matter
it has to be done.

Hirst was upset that Admiral Key was leaving, and praised him at the farewell dinner
arranged by the Officers of the College:

> Of me and my relations to him he spoke in the warmest and most gen-
> erous way and of the service I had rendered to him and the College he
> alluded to in terms of high recognition – higher than they merited. I
> waited until wine glasses had been replaced by coffee cups, until it was
> quite certain in short that Captain Molyneux had come to the end of his
> programme. I then asked him to let me say a word and I endeavoured to
> supplement his own too meagre account of Admiral Key's exertions in
> behalf of the College by a simple narration of facts which could have
> been known to no one present so completely as to myself. I spoke with all
> the warmth and earnestness I felt and what I said was warmly received by
> all. When I sat down I had contrary to my usual experience the satisfac-
> tion of feeling that every word I had uttered ought to have been uttered
> and that the praises I had bestowed on our parting President had been all
> abundantly merited.

Admiral Key later showed Hirst his last letter as President to the Admiralty:

> Its object was to put on record not merely his own estimate of my services but
> the fact that my duties had become much more onerous, my holidays much
> shorter than I was led to believe they would be on accepting the appointment.
> From first to last every act of Admiral Key's might be described as timely,
> just, kind, and graceful.

Hirst got on well with the five College Presidents under whom he served, but from
time to time, there were tensions between the College and the Admiralty. One of
these, in June 1876, concerned the timing of examinations on the training ship H.M.S.
Britannia. Hirst had proposed that these should commence as usual on the first Monday
in July, but when Admiral Fanshawe forwarded Hirst's letter to the Admiralty, he also
enclosed a letter from the Britannia's Captain proposing the second Monday.

> A day or two afterwards an official letter reached us intimating that Captain
> Graham's proposition was approved and remarking sharply "This is not
> the first time that the Examinations on the Britannia have been anticipated
> to meet the convenience of the Director of Studies and to the detriment of
> the cadets." In a very temperate reply to this letter I pointed out that the
> Admiralty had been entirely misinformed: that on every occasion the
> Captain of the Britannia had been consulted and the dates fixed in strict
> accordance with precedent. ... A few days afterwards a reply came not with-
> drawing the charge against me but clumsily admitting that the Admiralty
> had been "misled into the statement that the convenience of the *College* had
> been unduly consulted". The want of candour in thus substituting the word
> College in place of Director of Studies in the above retraction was more than
> I could bear and I replied that the charge against me personally must be
> withdrawn as distinctly as it was made otherwise it would be impossible for

me to maintain satisfactory relations either with the Admiralty or with the Captain of the Britannia with whom I was necessarily thrown into personal communication.

An important event took place on H.M.S. Britannia two years later when their Royal Highnesses the Prince and Princess of Wales arrived on board to present the prizes, in the presence of a large and distinguished crowd. Thomas Hirst detailed the work of the cadets, after which the Princess distributed the prizes, which ranged from a gold compass and a double aperture telescope to binoculars, despatch boxes, and books. The proceedings concluded with speeches by the Prince and by the First Lord of the Admiralty, the Rt. Hon. W. H. Smith[26], who invited everyone to show their thanks, 'as English boys always can exhibit their pleasure, by ringing and hearty cheering'.

Shortly after this, in October 1878, Hirst saw a less attractive side to the First Lord, who wrongly berated him for a statement he claimed that Hirst had made:

> It appears that he did not say that my statement was untrue (or rather he did not mean to say so, for say it he certainly did) but that it was *no longer* true. Of course I accepted the explanation. I took occasion to tell him however that my position, already difficult enough, would become untenable – that indeed I should not care to hold it – were there to be any loss of confidence in me on the part of the Admiralty. Up to the present time Mr Smith's attitude towards me and towards the College does not, I confess, give me any satisfaction. There is a state of tension between us which may at any time end in a rupture.

But there were compensations. In May 1881, along with other professors at the College, Hirst received a special decoration, given in recognition of the services rendered to a group of Chinese Officers who had studied at the College a year or two earlier:

> This day permission was given by Admiralty to myself and several Professors and Instructors at the College to accept the Chinese decorations and medals which have recently been forwarded to me through the Chinese mission at Paris. ... Mine is an order of the first class of the "precious star" and is decidedly curious.

7.6 Geometry

Although Hirst continually complained in his diaries about lack of time for his geometrical researches, he managed to complete several papers while at Greenwich.[27] Shortly after arriving at the College, he noted[28]:

> I commenced to-day writing an account of my Geometrical Researches on the Correlation of Two Planes. I wish very much before College work commences to put on record my two years' work on this subject, and the more so because in a posthumous paper of Clebsch's recently published in the *Mathematische Annalen* the subject is incidentally touched upon.

A *correlation between two planes*, sometimes called a *homography*, is a correspondence in which each point x (the *pole*) lying on a line X in one plane corresponds
uniquely to a line Y (the *polar*) passing through a point y in the other and *vice versa*;
the concept arose from some work of Chasles in the 1860s. The account was eventually finished and ready for the printers:

> Left my paper with Hodgson & Son to be printed. I propose to present it, in
> proof, to the Mathematical Society, and to let them have 500 copies for the
> Proceedings at cost price of paper and printing off; I bearing that expense of
> putting it into type.

Over the summer vacation of 1874, Hirst travelled on the Continent and visited
Rudolf Sturm[29] in Darmstadt:

> He received me with all the cordiality of a brother Geometer and I endeav-
> oured to describe to him my recent researches, and to induce him to assist me.

By now, Hirst had extended his researches to the more difficult topic of
correlations in three-dimensional space. On 12 November 1874, Hirst concluded his
two-year term as President of the London Mathematical Society, and his valedic-
tory Presidential address on correlation in space was described by his successor,
Henry Smith of Oxford, as containing 'a vast number of beautiful and interesting
results'.[30] But, as always, College work was preventing him from making the progress he desired:

> Mornings Geometry as usual. I am still working at Correlation in Space; but
> I make little progress. My time is so taken by my College work that I never
> can become immersed in my problems and as a consequence little way is
> made. I cling to it in hope of utilising more continuous leisure should it ever
> be granted me.

During the following Easter, he was still making limited progress:

> I have had ten days very concentrated work, though the difficulties to be
> overcome are so great that the results achieved is [sic] of no great visible
> magnitude. Until the midsummer holidays come I shall now be able to do
> but very little.

Over the summer of 1875, Hirst was back on the Continent, meeting Chasles,
Bertrand, Jordan, and Tchebichef in Paris and enjoying conversations with Sturm in
Germany.[31]

> After dinner Sturm and I had a pleasant walk through part of the Odenwald
> to Auerbach and much geometrical talk. He has begun to work at Correlative
> Bundles in view of my applying his results to my Researches on Correlation
> in Space. Hitherto he has confined his attention to Correlative Bundles which
> satisfy seven double conditions. I suggested he should consider simple condi-
> tions also.

In spite of insufficient time, his researches into correlation in space were indeed progressing well, proving to be 'rich in results both of interest and importance'. On receiving a letter from Sturm, he noted:

> His letters just now are of great value and interest to me, his results bear closely upon my own question of correlation in space. ... I wrote to Sturm giving him my solution of certain anomalies which have recently perplexed us both. I have almost invariably found that the resolution of anomalies leads to greater insight into the whole question under investigation and so it has been here.

In a correlation between two planes, two points (or lines) are *conjugate* if the polar (or pole) of either passes through (or lies on) the other. Hirst called the lines joining pairs of conjugate points a *complex* and investigated their coordinates.

By 1878, Hirst's researches were beginning to be appreciated. In Italy, a complex was named 'un complesso Hirstiano', while at home, his work came to be recognized in Cambridge:

> On the 8th I received the Diploma of Membership of the Cambridge Philosophical Society. I was gratified about a month ago to hear through Glaisher of my election. I may say that this is the first recognition I have ever received from any University in my native country.

But all too often, the tale was one of missed opportunities:

> The first meeting of the Math. Soc. took place on Nov. 11th. Cayley came to it and stopped with me. We were speaking of Cantor's paper on the cyclical self-corresponding points in two coincident planes between which a quadric relation exists. It has just appeared in the *Annale di Matematica*. I communicated precisely the same theorem to the British Association at Birmingham in 1865 but nothing was printed about it except the barest notice in the Proceedings. I showed Cayley my M.S. notes for that communication. He took them home with him and expressed an intention to write something about the matter. I shall be glad to be associated with a theorem which was always a pet of mine. As usual however I went on nursing my pet with the intention of allowing it to grow and develop itself more before I published it.

However, Hirst continued to have the occasional success. In 1881, his earlier contributions to hydrodynamics featured in the report from the British Association meeting in York:

> Some very elegant illustrations of Huygens' principle as applied to liquid waves have been given by Hirst. He has developed the differential equations for the lines of ripple, or lines of disturbance, which give small waves ... and has applied them particularly to the case where the centre of disturbance describes a circle uniformly on still water. The results are compared with experiment, with full agreement between theory and practice.

Most notably, in November 1883, shortly after leaving Greenwich, his geometrical investigations over many years were publicly recognized when the Royal Society presented him with its prestigious Royal Medal:

> Dinner at the X-Club. Huxley told me that the Council of the Royal Society had just awarded to me the Royal Medal. ... I was, of course, at the General Meeting of the Royal Society when I received my Royal Medal from Huxley, who addressed to me a few friendly words in addition to the formal ones of presentation.

7.7 Illness

Hirst's entire Greenwich career was dogged by ill health. Only a month after the college opened, he started to suffer from severe numbness in his legs which caused him to stay at home for over a month and work from there.[32]

> Every body, Admiral Key included, came to me and College matters went on smoothly enough. I made all arrangements for the Examinations which commenced Dec. 19th and ended the term. ... Gradually however sensation in the legs grew more natural and the hands grew worse. The little fingers first lost sensation and became stiff and useless. This gradually extended to other fingers and at last whenever I attempted to close my fist painful tingles crept up the sides of my fingers. ... wretched as the month was for me personally, the sympathy which my illness elicited from my best friends was in one sense a source of great consolation though its manifestation almost upset me sometimes.

The following July his health deteriorated:

> On attempting to come down stairs and go out violent pain came on. With difficulty I got up stairs on the first floor again and there flung myself on the bed. I could not lie still there but rolled on the floor writhing with pain.

It turned out to be an acute attack of kidney stones. Following other health problems, he felt the need for an extended period of change and rest, due to the 'melancholy break down in my health', and the College gave him several months leave. He travelled to Egypt, and then to Greece and Italy.

> I am no longer so distressed about my arms and legs though I notice little change as far as their control by the will is concerned. After walking a mile or two my right leg and arm are very shaky, they lose all precision of movement.

Later in the trip, other maladies began:

> I have been suffering for some days from 'Catarrh in the Intestines', an unpleasant and prostrating malady, which superposed as it is on my habitual

feebleness renders me very helpless. … During the night I had an attack of vomiting. This was the commencement of an attack of lassitude, flatulency and dyspepsia complicated with neuralgia in the tongue which went on increasing for a fortnight until it became really serious.

In July 1882, Hirst had an operation for a stomach tumour, performed by Sir Joseph Lister[33] using the new anaesthetic:

He made me cough frequently and grasped the tumour while I did so. At length he appeared to be satisfied and he proceeded. I remember nothing until I awoke in bed intensely miserable. I felt I was shrieking hysterically, gasping for breath and I know not what.

During the operation, Lister's hand had accidentally slipped and cut him internally. Hirst was required to lie motionless for a week and take great care for a considerable time afterwards.

Meanwhile, several personal misfortunes had come his way. His close friend Tyndall married, and henceforth they saw little of each other. His brother John was involved in a cab accident, John's wife left him, and later, when drunk, he fell down stairs and was found dead with a dislocated neck. Lilly, his niece, housekeeper and travelling companion, left him to get married, later producing a healthy baby son after a frightful 56-hour labour and dying shortly afterwards.

7.8 Hirst Leaves Greenwich

After ten years at Greenwich, partly for reasons of health, he decided to resign his position. In February 1883, he wrote[34]:

It was on the 13th that I first told Admiral Luard of my intention to retire from my present appointment at the end of the Session.

In July, the Officers of the Royal Naval College gave a farewell dinner in his honour, where his speech was well received and he was greeted with hearty cheers. Determining to put this part of his life behind him, he burned all his old manuscript mathematical books and lecture notes.

The only outstanding issue was his pension. Hirst received a telegram from the Admiralty stating that his pension was £394 per year instead of the £450 he had expected: since he had been promoted from Assistant Registrarship to his Greenwich post, he could not count his time at London University and also have extra years added on to his time at Greenwich. Admiral Key was consulted and made the following proposal, which Hirst accepted:

Instead of the 10 years at Greenwich with the 7 added on, but no University time, I should have 10 at Greenwich, 3 at the University and 5 added on, retrospectively, to the latter making altogether 18 years instead of 17 to be counted for pension; which would accordingly amount to £417.18 a year. This he added was the best that could be done for me.

Hirst's successor at Greenwich was William Niven,[35] whose duties commenced on 1 September 1883. Four days later, Hirst returned briefly to Greenwich to discuss the results of the recent examination for the rank of Lieutenant:

> I was no longer Director of Studies. It was Niven who signed the Examination reports. Truly I felt that my vocation was at an end.

After leaving Greenwich, Hirst continued to work on his geometry, producing three further papers. In 1890, after working on it for 16 years, he finished his work on the correlation of two spaces and symbolically burned his mathematical notebooks. He would write no more.

He spent his last years at his London clubs during the summer, and in the South of France in the winter. In September 1891, sensing the end was near, he travelled to Paris to bid goodbye to Anna for the last time. Four months later, a flu epidemic, one of the worst of the century, hit London. Thomas Archer Hirst, his resistance lowered by years of illness, quickly succumbed: he died on 16 February 1892 and was buried in Highgate Cemetery.

Acknowledgement

The authors are grateful to Tony Mann for obtaining much useful material on the Royal Naval College from the Public Records Office at Kew. The extracts from the Hirst diaries appear by courtesy of the Royal Institution.

8

A Professor at Greenwich: William Burnside and his Contributions to Mathematics

Peter M. Neumann

CONTENTS

8.1 William Burnside and his Work

From the late nineteenth century on there have been a number of distinguished mathematicians (among them T. A. A. Broadbent and L. M. Milne-Thompson) who worked at Royal Naval College, Greenwich (RNC). The most distinguished, however, in terms of impact on mathematics, and the one who is now best remembered for his original discoveries, was William Burnside (1852–1927).

Burnside served as professor at RNC from 1885 until 1919. His duties were focussed on teaching naval officers the basic mathematics needed for navigation, gunnery, and naval engineering. He taught a full load but had little administration imposed upon him, and although research was not part of his formal duties, he had plenty of time to think about mathematics beyond the needs of his teaching. For details of his life and work the reader is referred to the obituaries by Forsyth[1] and the articles by Everett, Mann and Young[2], and Hawkins.[3] Here is a brief chronology:

1852 William Burnside born (2 July) in London.

1858 Orphaned (that is, his father died).

Educated at Christ's Hospital School.

1871 Entered St John's College Cambridge; later migrated to Pembroke College.

1875 Fellow of Pembroke College (until 1886); spent ten years coaching rowing and mathematics.

1885 **Appointed professor at Royal Naval College, Greenwich**.

1886 Married.

1887 Start of his steady output of original discoveries in mathematics.

1893 Elected Fellow of the Royal Society; start of his commitment to the theory
of groups, which is what he is now remembered for.

1899 Awarded De Morgan Medal by the London Mathematical Society.

1906 President of the London Mathematical Society (until 1908).

1919 **Retired from Royal Naval College, Greenwich**.

1927 Died (21 August) aged 75.

Burnside's career exhibits features that are unusual amongst highly original and
productive mathematicians. His first article was not published until 1883 when he
was over thirty years old. On his appointment to the chair of mathematics at RNC
Greenwich in 1885 his output was still only three articles totalling a mere twelve
printed pages. The reputation that earned him the professorship must have been
based almost entirely on his work as a lecturer, mathematical coach, and examiner
in Cambridge (though it would be nice to think – even although there is no evidence
at all – that an appointing committee consisting of naval men would not hold a man's
predilection for rowing against him). His fourth paper was published in 1887, the year
after his marriage, and from then on he published an average of four items a year,
a good output for a mathematician. Over his lifetime he wrote approximately 160
papers and two books,[4] though the two editions of his book on group theory could
perhaps be counted as separate items since the second edition contains much material
that is not to be found in the first.[5]

Much of his work in the early years at Greenwich was on applied mathematics. He
wrote on Boltzmann's theory of gases, conceived as the statistical mechanics of huge
numbers of molecules moving freely subject to collisions between them. His paper,
although not uncontroversial, was generally well-received and stimulated useful discussion. Inspired perhaps by his observations of the river Cam as a rowing man, and
probably by the observed behaviour of the tidal wave that travelled round the world
after the explosion of Krakatoa in August 1883, he wrote two papers on hydrodynamics, the first on waves in deep water and the second on small waves essentially in shal-
low water. His studies in hydrodynamics and in electromagnetic theory were to lead
him to questions in pure mathematics, in the theory of elliptic functions; he also wrote
on a variety of technical problems in geometry and the theory of functions (higher
calculus). It was this work totalling just over thirty papers that earned him election to
Fellowship of the Royal Society in 1893. Late in his career, stimulated by military and
naval needs arising from the Great War of 1914–1918, he turned his attention to probability and statistics. His thoughts in this area were summarised in the book[6] that was
published posthumously.

The year 1893 in which he was elected to the Royal Society was highly significant
for Burnside's career also in another way. It was the year in which his first paper on
the theory of groups was written. His contributions in this area have lasted far better
than any of his other work. His name is attached to a number of concepts, theorems,
and problems, and it is for his work in group theory that he is now remembered and
revered. The second half of this paper will be an attempt to explain in a relatively non-
technical way what this subject is and what Burnside achieved in it.

8.2 The Theory of Groups

Groups enter science in a number of different ways, but their most important features are that they are abstract algebraic objects and that their principal function is to measure symmetry.

When we speak of measuring symmetry we have in mind formal symmetry such as that of a circular object, or a square, or a crystal. We measure their symmetry by considering their so-called *symmetry operations*. These are the things one could do, or imagine doing, to the object so as to leave it apparently unchanged. Obviously one of those things one could 'do' is leave it alone. The action of leaving something alone, that is, of doing nothing, will be denoted by I, and it is a symmetry operation that is always there (and, in spite of its apparent triviality, is very important). Often an object has many more symmetry operations however. For example, if a plain cylindrical vase standing on a horizontal table were to be rotated through any angle about its axis (the line perpendicular to the table and through the centre of its base), then it would remain apparently unchanged; anyone present whose eyes happened to have been closed while the vase was being turned could not detect that anything had occurred. Or think of a triangle ABC (Figure 8.1). If it is scalene (all three sides of different lengths) then all one can do to leave it apparently unchanged is leave it actually unchanged; that is, I is its only symmetry operation; if it is isosceles, so that sides AB, AC are the same length, then one could turn it over before returning it to its original position and it would be apparently unchanged, though vertices B and C had been interchanged; if it is equilateral then there are six operations that leave it apparently unchanged:

I: leave it alone.

R: rotate it through one-third of a turn clockwise, so that A, B, C move to the places originally occupied by B, C, A, respectively.

S: rotate it through two-thirds of a turn clockwise, so that A, B, C move to the places originally occupied by C, A, B, respectively (this is the same as rotating it through one-third of a turn anticlockwise).

U: turn it over the axis through A, so that A stays fixed and B, C move to the places originally occupied by C, B, respectively.

V: turn it over the axis through B, so that B stays fixed and A, C move to the places originally occupied by C, A, respectively.

W: turn it over the axis through C so that C stays fixed and A, B move to the places originally occupied by B, A, respectively.

The numerical facts that our plain circular vase has an infinite number of symmetry operations (the angle through which it may be turned could have any size), the scalene triangle has just one (do nothing), the isosceles triangle has two, and the equilateral triangle has six already provide some measure of the symmetry of these objects, and confirm in a formal way our instinctive observation that their symmetry properties are different. The mere number of symmetry operations of an object is, however, an inadequate measure of its symmetry.

Compare, for example, the equilateral triangle with the six-armed swastika (Figure 8.2). The latter also has six symmetry operations. They are I, R_1, R_2, R_3, R_4, and

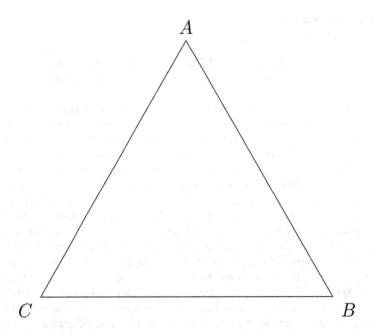

FIGURE 8.1 Triangle.

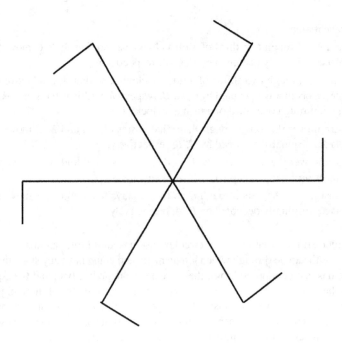

FIGURE 8.2 Six-armed Swastika.

R_5, where I is the operation of doing nothing, and R_k is an anticlockwise rotation through k-sixths of a full turn, that is through $(k \times 60)°$. Although both figures have precisely six formal symmetries, we feel instinctively that their symmetry is different. And of course that is true. But to understand the difference, and make it more precise, we need a more delicate measure of symmetry than merely the number of symmetry operations.

To measure symmetry more precisely we exploit the fact that the collection of symmetry operations of an object has structure. Namely, its symmetry operations can be 'multiplied'. The technical term is *composed*. If x and y are symmetry operations then their composite xy is the operation 'do x, then do y'. For example, a rotation through 20° composed with a rotation through 40° is simply a single rotation through 60°. The set (collection) of all symmetry operations of an object together with the rule for composing those operations forms a *group*, and it is this group which measures the symmetry of the object.

To see what this means examine the following tables which record the result of combining the symmetries of our equilateral triangle and of our six-armed swastika. The entry in the row labelled x and the column labelled y is the composite xy (Tables 8.1 and 8.2).

The tables look very different not merely because we have used different symbols to denote the symmetries of the two figures, but also because they really are fundamentally different. For example, if x is a symmetry of the triangle, then performing x twice or three times produces I, that is, $x^2 = I$ or $x^3 = I$, whereas there are symmetries x of the swastika (such as R_1 and R_5) for which $x^2 \neq I$ and $x^3 \neq I$. Another difference is that there are four symmetries x of the triangle such that $x^2 = I$, whereas there are only two such symmetries of the swastika. A third great difference is that the system of symmetries of the swastika is *commutative*, that is $xy = yx$ for any two

TABLE 8.1

Symmetries of the Triangle

	I	R	S	U	V	W
I	I	R	S	U	V	W
R	R	S	I	V	W	U
S	S	I	R	W	U	V
U	U	W	V	I	S	R
V	V	U	W	R	I	S
W	W	V	U	S	R	I

TABLE 8.2

Symmetries of the Swastika

	I	R_1	R_2	R_3	R_4	R_5
I	I	R_1	R_2	R_3	R_4	R_5
R_1	R_1	R_2	R_3	R_4	R_5	I
R_2	R_2	R_3	R_4	R_5	I	R_1
R_3	R_3	R_4	R_5	I	R_1	R_2
R_4	R_4	R_5	I	R_1	R_2	R_3
R_5	R_5	I	R_1	R_2	R_3	R_4

of its symmetries x, y, whereas there are symmetries of the triangle which do not commute: for example $RU = V$, whereas $UR = W$. Thus although they both have six symmetry operations, their actual symmetry is very different, and we can capture this difference in a formal way through comparison of their symmetry groups.

Let us focus, then, on groups, ignoring for the time being the objects whose symmetry they measure. The first ingredient of a group is a *set*. In the case of symmetry groups this is the set whose members are the symmetry operations of an object. For the equilateral triangle, for example, it is the set $\{I,R,S,U,V,W\}$; in the case of the six-armed swastika it is the set $\{I,R_1,R_2,R_3,R_4,R_5\}$. The second ingredient is a *binary operation* on the set. A binary operation is a recipe for combining any two members of a set to get a third. Usually we write something like $x \circ y$ for the result of combining members x and y; often this is abbreviated to $x y$. (Notice that we are now using the word 'operation' with two quite different meanings. Unfortunately this technical terminology is so well established in mathematics that we must live with it, but from the context it will always be clear whether what is meant is a symmetry operation or a binary operation.) Examples of binary operations are addition or multiplication of numbers; in the case of symmetry groups, it is the operation of composition of the symmetry operations. The third ingredient of a group is a distinguished member e of the set. It is known as the *identity*. In the case of symmetry groups this is the operation 'do nothing' that was denoted I above. For a group these three ingredients are required to satisfy certain conditions:

1. $(x y) z = x(y z)$ for any members x, y, z of the set. [Associativity]
2. $e x = x$ and $x e = x$ for any member x of the set. [Identity]
3. for every x in the set there exists y in the set such that $x y = e$ and $y x = e$. [Inverses]

From these conditions one can easily derive the facts that e is the only identity element of the group (that is, the only member of the group for which condition (2) holds) and that each element x has a unique inverse (that is, there is exactly one member y for which condition (3) holds). The inverse of x as given by condition (3) is usually denoted as x^{-1}. In fact the identity e and the inverses x^{-1} can be identified once the binary operation is known, so it is the binary operation which really determines the group. But the identity and inverses are so important that we find it best not to lose explicit sight of them.

Coming back to the question implicitly posed in the first paragraph of this section we now have an answer to the question what *is* a group? The answer is that it is a set G equipped with a binary operation and a distinguished element e satisfying conditions (1), (2), and (3) listed above.

The symmetry operations of an object certainly do form a group. Associativity of composition is simply the fact that $(x y)z$ is the result of performing the operation x, then the operation y, and then the operation z, as is $x(y z)$. The operation I, 'do nothing', is the identity element e, as is very easy to see. And for any symmetry operation x there is always an inverse x^{-1}, namely the operation 'undo x'. This group measures the symmetry of the object. And in a similar sense all groups measure symmetry. Although they do have some other interesting and important uses, this is the primary answer to the question what are groups *for*? They are for measuring symmetry.

8.3 Some Basic Theory of Groups

Most of Burnside's work was on finite groups, that is to say, groups that contain only finitely many members (like the symmetry groups of the triangle and the swastika considered above, and unlike the symmetry group of the circular vase). From now on therefore we will restrict attention to the finite case.

This is no place for a course on the theory of finite groups. Nevertheless, the reader who is interested in appreciating Burnside's place in science must be exposed to just a little of the basic theory. The following is intended therefore as an impressionistic guide to the main relevant points.

The first step in understanding groups is to realise that groups which are actually different may be essentially the same. Here is a very simple example. Take G_1 to be the symmetry group of a non-equilateral isosceles triangle ABC. As we noted above it consists of two symmetry operations. They are I (do nothing) and U, the operation of turning it over its axis through the vertex A, so that the base vertices B,C get interchanged. Now take G_2 to be the group consisting of the numbers 1 and -1, with ordinary multiplication of numbers as its binary operation. It is easy to check that this really is a group according to the definition. The multiplication tables of G_1 and G_2 are simply

G_1	I	U
I	I	U
U	U	I

and

G_2	1	-1
1	1	-1
-1	-1	1

Technically these are different groups because their sets of members are not the same. But in fact the difference is quite insignificant. If we identify I with 1 and U with -1, then we see that composition is the same. In general we would not want to distinguish groups whose multiplication tables (that is, whose binary operations) are the same except for a re-labelling or a re-naming of the members. Two groups with this property are said to be *isomorphic*, and for most purposes it is useful not to distinguish isomorphic groups. The group of symmetry operations of the equilateral triangle and the group of symmetry operations of the six-armed swastika considered above are *not* isomorphic – and it is this non-isomorphism which, to a great extent, captures the fact that these geometrical figures have very different symmetry.

The next step in understanding groups is to develop the idea of a *subgroup*. If G is a group, a subset H is said to be a subgroup if it is *closed* with respect to the basic structure (the binary operation, identity, inverses) of G. This means that for any members x,y of H the product $x y$ must also be a member of H, that e is a member of H, and that if x is a member of H, then x^{-1} must also be a member of H. When this happens H is a group in its own right. Note that $\{e\}$ and G are always subgroups. But most groups have non-trivial (other than $\{e\}$), proper (other than the whole group G) subgroups as well. For example, $\{I,R,S\}$, $\{I,U\}$, $\{I,V\}$, $\{I,W\}$ are subgroups of the symmetry group of the equilateral triangle. And $\{I,R_3\}$, $\{I,R_2,R_4\}$ are subgroups of the symmetry group of the six-armed swastika.

Suppose now that G is a group and H is a subgroup of it. The number of members of G is known as its *order* and denoted $|G|$; similarly of course, $|H|$ denotes the order, that is, the number of members, of H. Obviously $|H| \leq |G|$, but in fact a much stronger statement is true. Take an element x in G and focus for a moment on the set of elements xh as h ranges through H. This set is naturally denoted xH and is known as a *left coset* of H in G. It is a remarkable fact that two left cosets xH and yH are either the same (as sets of members of G) or completely different in the sense that they have no members in common—that is, they are 'disjoint'. Since our group G is supposed to be finite, there can only be finitely many of these cosets, say $x_1H,\dots,x_nH$. Every member of G lies in exactly one of them and each of them has the same number of members as H, that is, $|H|$. Therefore $|G| = n|H|$. The number n of left cosets is known as the *index* $|G:H|$ of H in G and so we see that $|G| = |G:H| \times |H|$. In particular, $|H|$ divides $|G|$. These facts are known as *Lagrange's Theorem*.

Given an element x of a group G we can investigate what happens when we form the powers $x, x^2, x^3, \dots, x^k, \dots$ of x. We find that there will always be values of k such that $x^k = e$. The smallest positive m such that $x^m = e$ is known as the *order* of x. Then the subset $\{e, x, x^2, x^3, \dots, x^{m-1}\}$ forms a subgroup of G. No two of the displayed elements coincide and so this subgroup has order m. From Lagrange's Theorem we know then that m divides $|G|$; that is, the order of any element divides the order of the group (a fact that is also known as Lagrange's Theorem).

Lagrange's Theorem(s) tell(s) us that the theory of finite groups should have quite a strongly arithmetical flavour. It is not so much the sheer size of a group G as the prime factorisation of its order $|G|$ which determines how complicated it can be. Another arithmetic theorem is *Cauchy's Theorem*, which states that if a prime number p divides $|G|$, then the group G must have an element of order p. A stronger result is *Sylow's Theorem*, part of which states that if a prime-power p^a divides $|G|$, then G will have a subgroup of order p^a. In fact, if p^α is the highest power of p that divides $|G|$, then the subgroups of order p^α are known as the *Sylow p-subgroups* of G, and they play an immensely important role in group theory.

For the discussion of Lagrange's Theorem above I introduced the left cosets xH of a subgroup H. One could equally well have worked with the right cosets Hx. Usually the collection of left cosets is different from the collection of right cosets. It can happen, however, that every right coset happens to be a left coset (and vice versa). In this case one finds that $xH = Hx$ for every member x of G. Subgroups for which this happens are said to be *normal*. In case H is a normal subgroup the set of cosets carries a natural binary operation which turns it into a group. This group is known as the *factor group* and is written G/H . Positive integers provide a very rough (but nevertheless helpful) analogy. Think of our group G as analogous to an integer n (say its order $|G|$) and a normal subgroup H as analogous to a divisor m (such as, by Lagrange's Theorem, its order $|H|$) of n. Then the quotient group is analogous to the quotient n/m.

This analogy, as I have said, is very rough. In particular, one of the reasons that basic group theory is much harder than basic number theory is that the analogy of multiplication (as the reverse of division) fails. If n is a number with a divisor m whose quotient is k, then we know that there is only one possibility for n; namely, n must be the number $m \times k$. But, given groups A and B it usually happens that there are many different (non-isomorphic) groups G that have a normal subgroup isomorphic to A with factor group isomorphic to B. Such groups are known as *extensions* of A by B, and what I am saying is that there may be many different such extensions. The symmetry groups of the equilateral triangle and the six-armed swastika again illustrate this phenomenon. Let G_1 and G_2 denote these respective groups. In G_1 we found the subgroup H_1 consisting of I, R, S; in G_2 we found the subgroup H_2 consisting of I, R_2, R_4. One checks easily that H_1 and H_2 are normal subgroups of G_1 and G_2, respectively; also that they are both isomorphic to the cyclic group C_3 consisting of $\{0,1,2\}$ with addition modulo 3, and that G_1/H_1 and G_2/H_2 are isomorphic to the cyclic group C_2 (consisting analogously of $\{0,1\}$ with addition modulo 2). Thus both of these symmetry groups are extensions of C_3 by C_2, and yet, as we have seen, they are very different groups. The problem of classifying all groups G which are extensions of a given group A by another given group B is known as the *extension problem*. It is a huge and sophisticated part of modern group theory. Interestingly, it is one of the areas of the subject to which Burnside contributed rather little.

Every group does have normal subgroups. For example the trivial subgroup $\{e\}$ is always normal. Or again, the whole group G is a normal subgroup of itself. A nontrivial group (that is a group with more than one member), in which there are no nontrivial proper normal subgroups, that is, a group in which the trivial group and the group itself are the *only* normal subgroups, is known as a *simple* group. Examples are the cyclic groups of prime order. These are the groups C_p, where p is a prime number, that consist of the integers $0, 1, \ldots, p-1$ equipped with the binary operation of addition modulo p. Continuing the crude analogy introduced above, simple groups are analogous to prime numbers. There are, however, simple groups of composite order that are very different from the cyclic groups C_p described a moment ago. There are infinitely many of them, and they provide much of the reason why group theory is an immensely sophisticated and exciting subject. There are just five of size less than 1,000, and their orders are 60, 168, 360, 504, and 660.

Let's take our crude analogy with numbers a little further. Start with our number n. If it is prime then record this fact and stop. If not then it will have a non-trivial proper divisor. That is, there will be a number m such that $1 < m < n$ and m is a divisor of n. Now we treat the numbers m and n/m in exactly the same way, and we continue. The process will end with a sequence of prime numbers. For example, if our starting number n is 60, we might factorise it as 6×10, then deal with 6 and 10, and end up with the sequence 2,3,5,2 of prime divisors of 60. A similar process can be applied to finite groups. Start with our group G. If it is simple then record this fact and stop. Otherwise there will be a non-trivial proper normal subgroup H. Then H and the factor group G/H are smaller groups than G, and we can treat them in exactly the same way. The outcome of the process will be a sequence of groups that have no non-trivial proper normal subgroups, that is simple groups. The simple groups that we reach in this way are known as the *composition factors* of the original group G. Unfortunately (or rather, very fortunately, because this gives rise to an extremely rich and interesting part of group theory) the original group G cannot be re-created from its composition factors in as simple a way as the integer n can be retrieved from its prime factors (simply by multiplying them together). The reason is that, as was discussed above, the extension problem is very difficult. Nevertheless the composition factors do carry a lot of information about the group G. Moreover, the crude analogy with numbers can be carried one step further. Remember that the prime factorisation of a number is unique except that the order of the factors may be changed. Thus, for example, $60 = 6.10 = 2.3.5.2$, as above, but also $60 = 4.15 = 2.2.3.5$. Whichever way we choose to break our number down, we will end up with the same prime numbers occurring with the same multiplicities. There is an important theorem, the *Jordan–Hölder Theorem*, which assures us that the analogous assertion is true of finite groups. The collection of simple groups that we reach is independent of the actual process, that is, independent of the choice of non-trivial proper normal subgroup used at each stage of the reduction. To summarise this paragraph, each finite group G has a unique collection of composition factors, which are finite simple groups, and the group G may be thought of as created from them by a series of extensions (in the technical sense of that word introduced above).

Thus there is quite a strong sense in which finite group theory may be thought of as having two complementary strands to it: first, find all the finite simple groups; then solve the extension problem so as to know how to find every group having a given collection of simple groups as its composition factors. Burnside contributed very much to the first part of this programme, rather little to the second.

A particularly important class of finite groups is the class of *soluble* groups. These are the groups all of whose composition factors are cyclic of prime order, the groups C_p mentioned above. Their original importance derived from the criterion for solubility of equations proved by Évariste Galois in about 1830 (first published in 1846), but they are also of great importance within the theory of finite groups.

There is just one further set of basic group-theoretic notions that we need in order to give some idea of Burnside's achievements. Think of a set X. A *permutation* of X is a rearrangement of its members. It is in fact a symmetry operation (in the sense discussed above) of X thought of as an object having no structure at all; that is, as a collection of elements which have nothing to do with each other except that they happen all to be members of X. Permutations can be composed, and we obtain the important group known as $\mathrm{Sym}(X)$ which measures the symmetry of a structureless object. If X has n members then this group has order $n!$, that is $1 \times 2 \times 3 \times \cdots \times n$. Subgroups of $\mathrm{Sym}(X)$ are known as *permutation groups*. The theory of permutation groups was quite highly developed and sophisticated already in Burnside's time, and I cannot hope to give the general reader much of a feeling for it here. It must suffice to say that there are important classes of such groups, such as *transitive* permutation groups and *primitive* permutation groups which play a major role in the theory. There are also the so-called *Frobenius groups*, given that name because in 1902 G. Frobenius found a proof of a conjecture about these groups that had attracted much attention over the previous ten years. They had been introduced by a French mathematician called Edmond Maillet in 1892. They lie far beyond the scope of this essay, but I feel I must mention them, if only in passing, because Burnside, who called them 'groups of degree n and class $n - 1$', made significant contributions to their study, and, in particular, he invented a tool called the *transfer* in this context.

8.4 Overview of Burnside's Contributions to the Theory of Groups

When Burnside became interested in the theory of groups it was a subject that was just under fifty years old. The basic facts listed above, Lagrange's Theorem, Cauchy's Theorem, Sylow's Theorem, and the Jordan–Hölder Theorem, had already been established, and many interesting examples of groups were quite well understood. Nevertheless group theory was still in its infancy, and Burnside was one of the mathematicians who helped turn it into a sophisticated twentieth-century subject. Roughly, what he worked on may be categorised (very approximately in chronological order) as

    i. structure of groups whose Sylow subgroups are cyclic [here Burnside's work overlapped considerably with work of others, such as Frobenius and Hölder]

   ii. structure of Frobenius groups

  iii. the search for simple groups

  iv. transfer theorems

   v. character theory and applications [a theory invented by Frobenius in 1896 which I cannot explain here]

  vi. structure of groups of odd order

 vii. transitive permutation groups of prime degree

viii. the $p^{\alpha}q^{\beta}$-Theorem

  ix. the 'Burnside Problem'

   x. structure of groups of prime-power order

  xi. other relatively minor matters.

In this chapter, which is intended for non-specialists, I cannot hope to explain all these achievements. Nor can I hope to explain properly how they inter-relate with the contributions of Burnside's contemporaries working in the same area, often on the same problems. I shall restrict myself therefore to just a few (namely (iii), (vi), (viii), and (ix)), of which the first three may be seen as different parts of the same project, chosen either because they are relatively easy to understand or because they were enormously influential. Readers who seek a fuller account are referred to the articles by myself, M. F. Newman, Ronald M. Solomon, and Charles W. Curtis.[7]

8.5 Burnside and the Search for Finite Simple Groups

The search for finite simple groups can be seen as having two parts: one searches for and catalogues as many of them as one can find, and then one seeks to establish that the lists are complete. Both parts require great effort, and to some extent they proceed simultaneously and interactively. In his book *Traité des substitutions* (1870)[8] Camille Jordan started the process by producing several lists of groups he could show to be simple. He had the cyclic groups C_p of prime order p, he had the alternating groups $\mathrm{Alt}(n)$, of order ½ n!, for $n \geq 5$ (these consist of the even permutations of a set of size n), and he had various groups that can be described using matrices.

In 1892 a German mathematician, Otto Hölder, started the other half of the process by showing that there are no simple groups of order ≤ 200 other than those that had already been discovered. Several people continued this work, and, in particular, in 1895 Burnside extended it to cover all orders up to 1092. He also contributed theorems based on the arithmetical structure of the order of a group. For example, he and Frobenius classified (more or less independently) the simple groups of order n, where n is a product of up to five prime numbers. And he proved that if p is the smallest prime divisor of the composite number n, and n is the order of a simple group, then either p^3 divides n or p must be 2 and n must be a multiple of 12. Already in 1897, at the end of his monograph,[9] he had the insight to write

> No simple group of odd order is at present known to exist. An investigation as to the existence or non-existence of such groups would undoubtedly lead [...] to results of importance [...].

In 1900 Burnside started a deep study of groups of odd order, and by 1911, when he published the second edition of his monograph, he was able to list many suggestive facts about them which led him to write

> The contrast that these results shew between groups of odd and of even order suggests inevitably that simple groups of odd order do not exist.

The non-existence of simple groups of odd composite order became known as 'Burnside's Conjecture'. It is equivalent to the proposition that all groups of odd order are soluble, and was proved in a very famous paper[9] published in 1963 by Walter Feit and John G. Thompson.

The first of the quotations above, the one from the 1897 edition of Burnside's monograph, continues:

> Also, there is no known simple group whose order contains fewer than three different primes. [...] Investigation in this direction is also likely to lead to results of interest and importance.

The issue here is this. Suppose the order of a group G is of the form p^α, where p is prime and $\alpha > 1$. It had been shown by L. Sylow in 1872 that then G must be soluble. What if $|G| = p^\alpha q^\beta$, where p,q are distinct prime numbers and $\alpha \geq 1$, $\beta \geq 1$? By the time Burnside wrote his book it was known in many special cases that G would have to be soluble. More cases were treated in the years following the publication of the book. Then finally in 1904 Burnside found a very beautiful proof of the general theorem that such a group cannot be simple. It is an easy consequence that such a group is always soluble. His proof used the theory of group characters that had been invented by Frobenius in 1896, together with facts from algebraic number theory. In recognition of his achievement the result is widely known as *Burnside's $p^\alpha q^\beta$-Theorem*.

The search for finite simple groups continued throughout the twentieth century. A number of new families of simple groups were discovered in the 1950s and early 1960s, and a number of so-called 'sporadic' simple groups were discovered in the 1960s and 1970s. At the same time, techniques for proving that all simple groups were amongst those that had already been identified were becoming more and more sophisticated. By 1980 it was announced that the search was complete, and although some mistakes and oversights still had to be repaired in the following twenty years, most algebraists are now confident that the classification of the finite simple groups really has been achieved. Burnside, with his conjecture on groups of odd order, with his $p^{\alpha}q^{\beta}$-Theorem, and a number of other less easily describable contributions, laid many of the foundations for this remarkable project.

8.6 The Burnside Problem

In 1902 Burnside published a paper entitled 'On an unsettled question in the theory of discontinuous groups'. The question that he asked was about what are called *finitely generated* groups. Suppose we have members $g_1, g_2, ..., g_m$ of a group G. We can consider the collection $\langle g_1, g_2, ..., g_m \rangle$ of all those members of G that can be obtained by multiplying (composing) the elements g_i and their inverses again and again until no new members of G appear. That is, $\langle g_1, g_2, ..., g_m \rangle$ consists of the elements g_i themselves, their inverses g_i^{-1}, products $g_i^{\pm 1} g_j^{\pm 1}$, products $g_i^{\pm 1} g_j^{\pm 1} g_k^{\pm 1}$, and so on. This is a subgroup of G, and in fact it is the smallest subgroup of G that contains all the elements $g_1, ..., g_m$. If $\langle g_1, g_2, ..., g_m \rangle = G$, that is, if every member of G can be obtained by multiplying together some of the given elements and their inverses in some order, using one or more of them several times if necessary, then we say that the set $\{g_1, ..., g_m\}$ *generates* G, and (given that this set is finite, as the notation is intended to suggest) that G is a *finitely generated* group.

Burnside's question was this. Suppose that the group G is finitely generated. Suppose moreover that every member of G has finite order, that is, for every g in G, there exists $n > 0$ such that $g^n = e$. Must G then be finite? In due course this came to be known as the General Burnside Problem. It was answered negatively by Golod in 1964, and now there are many constructions known that produce finitely generated groups all of whose members have finite order, but which themselves are infinite.

Burnside moved on immediately, however, to what he called 'a special form of this question'. For a positive integer n, a group G is said to be of *exponent n* (strictly of *exponent dividing n*) if $g^n = e$ for every member g of G. In a group of exponent n every element has finite order, and moreover that order divides n. What came to be known as *The Burnside Problem* was this: if a group G is finitely generated and of exponent n, must it be finite?

This question has turned out to be extraordinarily productive of good mathematics. There is a group, known as $B(m, n)$ which is in a strong sense the largest group of exponent n that can be generated by m elements. The question therefore can be reformulated as: is the group $B(m, n)$ finite? But then it is natural to add a rider: if so, what is its order?

For $n = 1$ the question has a trivial answer. For $m = 1$ the answer is almost as easy— the group $B(1, n)$ is a cyclic group of order n. For $n = 2$ it was already well known in Burnside's day that $B(m, 2)$ is finite and of order 2^m. In his paper Burnside treated the groups $B(m, 3)$ and $B(2, 4)$ and showed that these were indeed finite.

It is now known that if n is 1, 2, 3, 4, or 6 then the groups $B(m, n)$ are finite for all m. It is also known that if n is large enough (for example, $n \geq 8000$), then $B(m, n)$ is infinite for $m \geq 2$. There are, however, still huge gaps in our knowledge. For example, we do not yet know whether or not the groups $B(m, 5)$ are finite for $m \geq 2$.

Over the years the Burnside Problem has stimulated many advances in group theory. For example, a variant known as the Restricted Burnside Problem, which asks whether there is a largest finite group of exponent n that can be generated by m elements, has led to what are called linear methods in group theory. One of these, the so-called Lie-ring method, is used to solve a very wide range of problems about groups of prime-power order. Another introduced by Philip Hall and Graham Higman in a paper on the Restricted Burnside Problem published in 1956, developed tools which have had an enormous impact on the search for finite simple groups. And in the course of all this work the Restricted Burnside Problem has been solved affirmatively by Efim Zelmanov, although there is still much that is unknown about the largest finite group $R(m,n)$ of exponent n that can be generated by m elements.

8.7 Conclusion

As a research mathematician William Burnside was unusual in many ways. One has already been commented on – the lateness of his start. Another is what one might describe as his professional loneliness. Of course he was not lonely in the usual sense. As an active member of the Royal Society and the London Mathematical Society he had regular contact with creative mathematicians of the highest ability. And at least in his work on applied topics, analysis, and geometry he could, and did, correspond with like-minded Cambridge friends. But he was alone in two ways. First, working in southeast London he was not in daily contact with a group of research-active colleagues. It is common nowadays to think of research groups as necessarily requiring a 'critical mass' of researchers. Of course that is a modern phenomenon, and to compare Burnside's environment with modern expectations is unreasonable. Nevertheless, even for his time Burnside seems to have been more of a loner and more self-sufficient than most.

The second aspect of his lonesomeness was in his choice of major topic for study. There were many of his contemporaries in other countries who wrote about group theory – F. Klein, R. Fricke, O. Hölder, G. Frobenius, I. Schur, A. Loewy in Germany; G. Frattini in Italy; C. Jordan, E. Maillet, Élie Cartan in France; F. N. Cole, E. H. Moore, L. E. Dickson, G. A. Miller, H. Maschke, H. L. Rietz in America, for example – but he was the only British mathematician with an interest in the subject. There is some evidence that he felt this isolation a little. In the preface to the 1897 edition of his book[10] Burnside wrote:

> The subject is one which has hitherto attracted but little attention in this country; it will afford me much satisfaction if, by means of this book, I shall succeed in arousing interest among English mathematicians in a branch of pure mathematics which becomes the more fascinating the more it is studied.

And the following is from the published version of the Presidential Address he gave to the London Mathematical Society on 12 November 1908 when he demitted office:[11]

> It is undoubtedly the fact that the theory of groups of finite order has failed, so far, to arouse the interest of any but a very small number of English mathematicians; and this want of interest in England, as compared with the amount of attention devoted to the subject both on the Continent and in America, appears to me very remarkable.

But these passages seem quite detached and dispassionate in tone, and it seems that he was not particularly unhappy in his isolation.

Perhaps most remarkable about Burnside is how many of his ideas and how much of his mathematics survived in the twentieth century. Arthur Cayley was thirty years older and died in 1895. Therefore he was not exactly a contemporary. Comparison is nevertheless revealing. Cayley was far more famous and celebrated in his lifetime. But by the third decade of the twentieth century almost all of Cayley's work had been superseded. Burnside's work in applied mathematics, analysis, geometry, probability and statistics suffered the same fate. But his techniques, his theorems, and his ideas about group theory remained influential throughout the twentieth century, and many are still in use today. Although a late starter and a lonely researcher, his contributions to group theory earn him his place up there in the pantheon of great mathematicians.

Acknowledgement

It is a pleasure to acknowledge with warm thanks that I have benefitted considerably from a thoughtfully critical reading by Dr J. A. Stedall of an early draft of this chapter.

9

The Nautical Almanac Office and L.J. Comrie: Mechanising Mathematical Table-Making at Greenwich

Mary Croarken

CONTENTS

9.1 Introduction

The Nautical Almanac Office (NAO) had been set up in 1832 in response to vigorous calls for reform from British Astronomers.[1] In 1922 it moved into rooms located in the King Charles Building at the Royal Naval College in Greenwich,[2] where the Office remained until 1939 when it moved to Devonport House in King William Walk, Greenwich before being evacuated to Bath during World War Two. During that time Leslie John Comrie (1893–1950) joined the Office and completely revolutionised the computing methods employed there. He installed a range of calculating and accounting machines and applied them to the table-making process, developing new numerical methods to make the best use of the characteristics of each machine. Through his work on machine computing methods at the NAO, his publications, and involvement with other mathematical table-making projects, Comrie was active in' spreading the gospel of mechanised computation'[3] outside of the NAO and influencing a whole

FIGURE 9.1   Leslie John Comrie. (Courtesy J. K. Comrie.)

generation of mathematical table-makers and scientific computers from his base at Greenwich (Figure 9.1).

9.2 The Pre-1925 Nautical Almanac Office

From its inception in 1767 to the early nineteenth century, the *Nautical Almanac* was held in very high regard for the content and reliability of its tables. Following Maskelyne's death in 1811, the tight supervision Maskelyne had imposed on his computers was loosened and standards fell. Maskelyne's successor as Astronomer Royal,

John Pond, took on the responsibility for annually producing the *Nautical Almanac* but had little interest in computational astronomy or navigation, and left coordinating the work of the computers to his relatively inexperienced comparer Thomas Brown.[4] Coupled with the retirement or death of several of Maskelyne's experienced computers, Pond's lack of control over the quality of the computing soon led to a rapid drop in the accuracy of the tables of the *Nautical Almanac*, and questions were asked in Parliament.[5] To solve the problem the well-known scientist and translator of the Rosetta Stone, Thomas Young (1773–1829), then secretary to the Board of Longitude, was appointed in 1818 as the first Superintendent of the Nautical Almanac. Young maintained the distributed network of computers that Maskelyne had put in place but personally took over the allocation of computing work, increased the supervision of the computers' work, and ensured that the checking and comparing work was done.

Young was successful in restoring the accuracy of the tables[6], but during the 1820s there were calls for further reform of the *Nautical Almanac*, largely from the astronomical community. Some British astronomers felt that the *Nautical Almanac* should cater for observational astronomers as well as navigators, but Young felt that the principal purpose of the *Nautical Almanac* was navigation and this was the product that the Admiralty were paying for. Consequently he resisted all calls for reform. When Young died in 1829, the Admiralty asked the Astronomical Society to report on the reforms they felt were needed. The report was published in 1831[7] and led to the creation of a NAO with a full-time Superintendent and a new staff based in London. William Stratford (1790–1853) was appointed as Superintendent and set up the NAO in 1832 in offices off the Gray's Inn Road in London. In doing so he disbanded the distributed network of computers and centralised the computing of the *Nautical Almanac* in London; only one of the existing Nautical Almanac computers came to London to remain in post.[8]

In 1922 the Admiralty, in a cost-cutting move, allocated the NAO three large rooms in the south-west corner of the King Charles Building at the Royal Naval College, it being cheaper to house the Office in an Admiralty building rather than paying for commercial accommodation.[9] In addition close proximity to the Royal Greenwich Observatory was appropriate if not strictly necessary and, while it probably had no bearing on the Admiralty decision, it was closer to the Blackheath home of the then Superintendent Philip Cowell (Figure 9.2).

Cowell (1870–1949) was Superintendent of the NAO from 1910 to 1930. Before taking up the post, Cowell had earned a reputation for his work in celestial mechanics, particularly for his work on the motion of the Moon and the calculation of the orbit of Halley's Comet in preparation for its 1910 return. When Cowell took up the post of Superintendent of the NAO, he had hoped to develop the NAO into a research centre for dynamical astronomy, but the Admiralty provided no support for this and Cowell had to settle for managing the traditional work of the Office.[10] Cowell's experience of computational work with the Moon and Halley's Comet left him well placed to improve and update the computing methods used in the Office. Most of the tables were compiled by hand using logarithm tables, and checked using differencing and an elaborate system of proof-reading.[11] Cowell organised the work of the office around a mix of skilled and experienced staff, many of whom had been working either in the NAO or at the Royal Observatory for many years on temporary contracts and worked outside the office, and less experienced temporary boys and men based at the NAO.[12]

FIGURE 9.2   King Charles Building, Royal Naval College, Greenwich. Site of the NAO 1922–1939.

Cowell also encouraged his staff to develop the computing methods used in the Office. One of the most well-known developments was by T.C. Hudson, an NAO Assistant, who devised a graphical method of finding the last digit of the second differences of functions. From the end figure the full second difference could be deduced and could then be used to interpolate between previously computed pivotal values of a function. To further improve this technique Hudson introduced into the Office a Burroughs Adding Machine that had been mechanically modified to operate in either sexagesimal or decimal, and Hudson used it to interpolate tables of the positions of the Moon, Venus, and Mars using the last digits of the second differences.[13] Comrie later recalled that Hudson's work inspired many of his own ideas and provided a platform on which he could revolutionise the computing methods of the Office (Figure 9.3).[14]

9.3 Comrie's Early Career

Comrie was born in Pukekoke, New Zealand, in 1893. While studying chemistry at the University of Auckland, Comrie developed an interest in astronomy. Despite considerable deafness, Comrie served in France with the New Zealand Expeditionary Force where he lost a leg in 1918. While recuperating in Britain Comrie took advantage of educational courses put on by University College, London, and had his first lesson on a Brunsviga calculating machine from the mathematical statistician Karl Pearson (1857–1936) on Armistice Day.[15] This

FIGURE 9.3 Burroughs Adding Machine c.1926. (Science Museum 10459381 and 303411, courtesy of the Science and Society Picture Library.)

meeting with Pearson, who used calculating machines extensively in the Biometric Laboratory at University College, London, was to revolutionise Comrie's career. At the end of the war Comrie enrolled as a postgraduate student at Cambridge University studying planetary occultations and methods for their computation, for which he was awarded his PhD in 1923.

While at Cambridge Comrie got his first experience of organising a large-scale computing project when he was invited to direct the computing section of the British Astronomical Association, supervising eight volunteers to work on reductions of observations.[16]Astronomical observations are usually recorded as four data points, height of the object above the horizon (altitude), direction in the sky (azimuth), time of day, and date. Reduction is series of calculations which converts (i.e. reduces) the raw data to a pair of fixed numbers (the right ascension and declination) which represents the position of the object in the celestial sphere and was the form that astronomers used for research. This project proved to be the start of Comrie's career in scientific computing. On leaving Cambridge Comrie took up lecturing posts in the United States in astronomy departments first at Swarthmore College, Philadelphia, and then at Northwestern University in Illinois. At both institutions Comrie introduced practical computing courses into the astronomy curriculum, and at the same time he began to publish articles on calculating machines and their application to scientific computing.[17]

Comrie returned to the UK in October 1925 to take up a post as Assistant in the NAO. His knowledge and enthusiasm for computational astronomy made him the perfect candidate for the post. Four months later he was promoted to Deputy Superintendent and 4 years later promoted to Superintendent on Cowell's retirement in 1930. Almost as soon as he arrived in Greenwich Comrie began to change the computation methods of the Office, and this must have been a factor in his appointment as his expertise was already growing in this area. While Cowell himself is reputed never to have used a calculating machine[18], he did have the foresight to see that they could and should be applied to the work of the NAO. He had encouraged Hudson and later supported Comrie by applying for funding to install computing machines at the NAO and by supporting Comrie's trips to further his knowledge of calculating machines.[19]

9.4 The Introduction of New Computing Methods into the NAO

On his appointment to the NAO Comrie immediately began to install calculating machines to expand upon and complement the Burroughs Adding Machine and Leyton Arithmometer already installed. By 1927 he had added an electrically powered Monroe, a Nova-Brunsviga and a Comptometer to the Office and by 1930s had an electric Monroe and a Hamann Selecta. Comrie built up good working relationships with calculating machine agents and manufacturers, and as machines came onto the market, he obtained them for trial in the office.[20]

Introducing calculating machines meant that Comrie had also to adapt existing computing methods to take advantage of the time- and labour-saving potential of the machines. At its most basic this meant working with tables giving the natural values of trigonometric functions rather than logarithmic values. As early as July 1926 Comrie had a programme of work underway which used the Burroughs Adding Machine to compute a seven-figure table of natural values of six trigonometrical functions for every second of time, a complete set of four-figure tables of natural trigonometrical functions, squares, cubes, roots, reciprocals, etc. and specific portions of natural trigonometrical functions to every second of arc. It was necessary to compute these in-house due to the poor reliability, when tested by Comrie and his staff, of existing published tables of natural trigonometrical functions.[21]

Whereas Cowell had used external workers paid on piece rates for a significant portion of the Office's work, Comrie's application of calculating machines to the work meant that he also had to change the make-up of the workforce. Rather than skilled male arithmeticians (many of whom were at or near retiring age) Comrie required and obtained calculating machine operators. Initially he trained some of the younger assistants to do the routine work under his direction but later hired staff with their potential as machine operators in mind. Many of these operators were women, and Comrie trained many of them personally. His actions in relation to hiring the staff he needed, particularly his desire to employ married women which was against Civil Service policy, often brought him into conflict with the Admiralty, but for the most part he managed by using a mix of permanent and temporary staff.[22]

9.5 End-Figure Interpolation

Having provided the basic tools for machine computation, Comrie began to further develop computing techniques for use with calculating machines. Interpolation was a fundamental part of the Office's work. In the mid-1920s the Office commonly used both Lagrangian interpolation and the building up of interpolates by continuous summation of finite differences. Comrie thought Lagrangian interpolation overly laborious and identified that it was likely to conceal any errors there might be in the original data,[23] and so concentrated his efforts in improving techniques for continuous summation of finite differences.

Comrie identified that Hudson's graphical method of finding the end figures of second differences had the significant disadvantage that the technique required a high degree of skill and could not be converted into a routine process to be carried out by junior staff. Using Hudson's method as a starting point Comrie developed a numerical method of deriving the end figures of second differences. For functions where the third difference was constant, knowledge of the end figures of the second differences meant that the whole second difference could be inferred and therefore the first differences and the interpolates could be calculated. In 1928 Comrie demonstrated how pre-computed tables of end figures of the terms of the Bessel's interpolation formula could be easily used to determine the end figures of second differences.[24] The pre-computed tables were published as an appendix to the 1931 *Nautical Almanac* and reprinted in 1936 as the pamphlet *Interpolation and Allied Tables* which became a classic in the field.[25] The end-figure method of interpolation required only simple table look-ups and simple arithmetic, and became routine in the NAO, but this was only the first stage – the function had still to be built up using the end figures (Figure 9.4).

The end-figure method of interpolation was applied to preparing the tables for the right ascension and apparent declination of the Moon for every hour of every day. The Moon's positions at noon and midnight were calculated individually and differenced to check for errors. The differences obtained from the table of 12-hourly positions were then used as arguments for Comrie's end-figure interpolation tables from which the first differences and interpolates could be derived in order to obtain the Moon's hourly positions.

9.6 Sub-Tabulation from Second Differences

Hudson had been using the Burroughs Adding Machine to build up a table from the end figure of the second differences for over 10 years before Comrie joined the Office. While the Burroughs machine of 1911 had the advantage of producing printed output, it had the disadvantage that for each table the computation, and the paper on which the outputs were printed, had to pass through the machine twice because the machine only had one accumulating register. First time through the machine computed the sequence of first differences from the end figures of the second difference, and second time through the machine was used to compute the function from the first differences with the operator having to input the first differences onto the machine.

By the mid-1920s calculating machine technology had advanced. In 1928 Comrie installed a Brunsviga-Dupla calculating machine that could be applied to building up

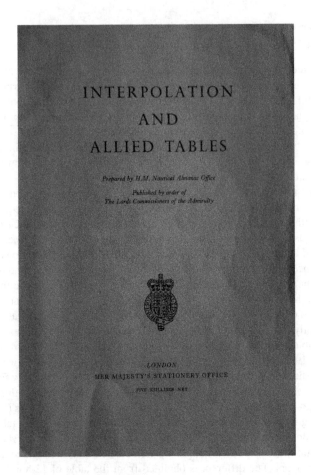

INTERPOLATION
AND
ALLIED TABLES

Prepared by H.M. Nautical Almanac Office

Published by order of
The Lords Commissioners of the Admiralty

LONDON
HER MAJESTY'S STATIONERY OFFICE
FIVE SHILLINGS NET

FIGURE 9.4   Interpolation and Allied Tables.

a table from its second differences.[26] Unlike earlier models, the Brunsviga-Dupla calculating machine had two product registers and had the facility that numbers could be input directly into one of those registers via thumb wheels at the front of the machine without having to disturb the value set on the setting levers. The machine could therefore be set up with the original function value in one product register, the initial first difference on the setting levers, and the second differences input each time via the thumb wheels with the interpolate being accumulated in the other product register. For each computation therefore only one passage of the machine was needed.

The drawback of the Brunsviga-Dupla was that it did not print. Errors could be introduced when the operator copied down the result and there was no record of the computation to check for errors. In 1929 a new Burroughs machine, the Class 11 Adding and Listing Machine, was introduced onto the market. Burroughs machines were manufactured primarily for the accounting market where their printing facilities were a crucial aspect of the audit trail – precisely what Comrie required in the NAO. Comrie was quick to experiment with and adopt the machine at the NAO.[27] The Burroughs Class 11 not only had two registers in which the function and the first differences could be accumulated when the second difference was entered via the keyboard, but also had control rods on the printing carriage which allowed the machine to be 'programmed'. Appropriate use of the control rods allowed several operations to take place during one cycle of the machine. In addition the results were printed across the page and the output of the machines used directly as printer's copy. The whole process, apart from entering the second differences, was now automated.

9.7 Punched Card Machines

While developing better, more efficient, ways of interpolation had a significant impact on the work of the NAO, particularly for computing tables of the positions of the planets or the declination of the Sun, Comrie's most radical computing revolution concerned the complete revision of the way in which the tables giving the position of the Moon for every noon and midnight were computed. The method of calculating the Moon's position in use at the NAO in the 1920s was based on theoretical and computational work undertaken by Ernest Brown (1866–1938) at the turn of the century. The individual positions were derived by Fourier synthesis involving the summation of a number of harmonic terms given in Brown's *Tables of the Motion of the Moon*.[28]

Brown's *Tables* consisted of 108 individual tables covering 660 pages over three volumes. Each table represented the harmonic terms required for specific periods of the Moon's motion. The computers had to extract the necessary 8- to 12-figure terms from the tables and sum them together. While the computation was conceptually easy to perform, it was laborious, and the constant copying down of table entries meant that it was open to error. The task kept two NAO computers fully employed all year round.

In 1929 Comrie automated this Fourier synthesis process using Hollerith punched card machines.[29] These were large electromechanical tabulating machines used widely for commercial data processing and accounting purposes. Data was punched as holes onto 45 column cards and 'read' by the closure of electrical circuits through the holes in the card. Cards could be sorted using a sorting machine and calculations performed using the tabulator. The tabulator had several accumulating registers and was set up using a plug board that physically connected the required registers with particular columns in the punched cards to give the required result. The same sequence of instructions was executed for each passage of a card through the machine although there was the possibility of introducing discontinuities that allowed the machine to stop for operator intervention. A printing mechanism was also attached to the machine so that totals produced by the machine could be produced in hard copy (Figure 9.5).

In the late 1920s Comrie began to see how punched card machines could be used to automate the calculations relating to the Moon's position. In May 1928 Cowell put in an official request to the Admiralty for funding to hire the machines for a limited period, and in December the Admiralty finally agreed.[30] The estimated cost would be £443 for the machines and cards plus staffing costs for machine operators. During 1928, while waiting for the final authority to come from the Admiralty, Comrie prepared for the work. First he organised for all of the entries in Brown's *Tables* to be transcribed onto punched cards preserving the order of the tables; approximately half a million cards were prepared over a 6-month period. Second he negotiated suitable space in the Apparatus Room of the Royal Naval College Engineering Laboratory to house the machines and convert the electrical wiring to the appropriate standard.

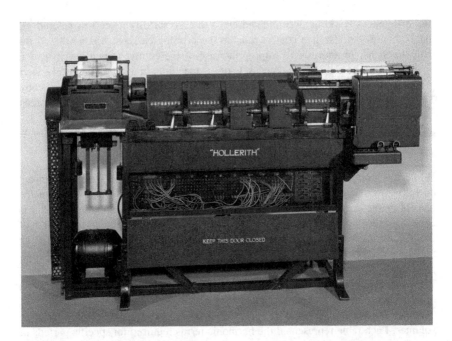

FIGURE 9.5 Hollerith Tabulating Machine c.1932. (Science Museum 10459381 and 303411, courtesy of the Science and Society Picture Library.)

In 1929 Comrie finally obtained a 7-month lease (rather than the 6 months costed by the Admiralty) on a set of Hollerith punched card machines, and the computations began. The process was simple. The operators selected the required cards by hand much in the same way as the values had been extracted from Brown's original *Tables*, and these were sorted to bring the necessary harmonic terms together for sequential dates and then passed through the tabulator. The resulting positions were printed, and the cards resorted into their original sequences and refiled in the main stock of cards. In the 7 months the Hollerith machines were installed at Greenwich, Comrie's staff completed the calculations of the positions of the Moon up to 1950 and partially completed the work up to 2000, at a total cost of £1,500 in terms of staff time and machine hire. To have undertaken the same work without the machines would have cost £6,000. The significance of Comrie's application of Hollerith punched card machines to the motion of the Moon computation was that it not only improved the techniques in the NAO but also signalled one of the first scientific applications of punched card machines. His work was to influence others all over the world.[31]

Despite no longer having the resources to maintain a punched card installation at the Royal Naval College, Comrie continued to employ a punched card machine operator and to apply punched card techniques to the work of the office. He achieved this by using spare capacity, when available, at other Hollerith installations located around London such as those at Imperial College and H.M. Stationery Office.[32] A plug board containing the machine set-up was prepared at the NAO, and the necessary cards were punched and selected. All sorting and tabulation was done at the remote site. As well as using punched cards to prepare the tables of the positions of the Moon, Comrie also applied them to preparing tables of the position of the sun and the planetary coordinates for the period 1940–1960, and used them for end-figure interpolation.

9.8 The National Accounting Machine as a Difference Engine

Checking the accuracy of tables by differencing had been a technique used with Nautical Almanac tables since 1767 (see Chapter 2). Originally Maskelyne had used a mixture of re-computation and differencing to check the tables, but by the 1920s differencing was the standard method of checking the accuracy of computed tables. While Comrie had applied several machines to subtabulation from second differences, when it came to using differences to check tables, two-register machines would not suffice as for most functions differences to four or six places were required to adequately detect errors. By the early 1930s Burroughs had brought out a six-register machine, but all inter-register transfers had to go through a single 'cross-footing' register instead of being transmitted directly and it was therefore not suitable as a difference machine.

In 1931 the National Accounting Machine Class 3000 (initially sold in the UK under the trade name of Ellis) came onto the market. It had six accumulating registers into which numbers could be added from the keyboard or from any of the other registers. The destination of inter-register transfers was controlled by stops on the printer carriage of the machine, and these could be arranged in sequence so that a function (or table) could be taken to fifth differences across one line before repeating the cycle, thus allowing its application to differencing.[33] Conversely the machine could also be used to build up a function from sixth differences set on to the keyboard – that is operated as a Babbage Difference Engine. The dual functionality of the machine made it an almost essential item for table-making in the mid-1930s.

When the machine was first available, the cost of the full six-register machine (almost £600) was beyond the limited NAO budget and the Admiralty had already spent almost £2,000 on calculating machines in recent years. Instead Comrie was instrumental in encouraging the British Association Mathematical Tables Committee (see below) to purchase a National to assist in their own programme of table-making. The machine was housed at the NAO and, when not in use on Mathematical Tables Committee work, Comrie could apply it to the NAO or other work. The machine was a decimal one, so it was not ideal for the largely sexagesimal work of the office, but by January 1934 the purchase of a sexagesimal National machine was approved by the Admiralty and it was much used in the Office.[34]

9.9 Comrie's Wider Table-Making Reputation

The efficiencies which Comrie introduced into the NAO by reforming the computing techniques used meant that the routine work of the office was well ahead of schedule.[35] As a consequence Comrie was able to use the facilities of the NAO to undertake other work and to build a reputation for himself as an internationally renowned table-maker and the NAO as an efficient and effective computing centre. He also attended international astronomy meetings and corresponded with table-makers all over the world.[36]

While reforming the computing work of the NAO, Comrie ensured that results of his work were published in a range of journals covering astronomy, accounting and statistics. Through his publications Comrie made sure that others heard of and had access to the machine computing techniques he developed. He gave talks to many organisations including the Royal Astronomical Society, the Royal Statistical Society, and the Office Machinery User's Association, and Manchester and Liverpool Universities, and spent some summers lecturing on machine computation at, for example, Imperial College (1929), University of California (1932) and University College, London (1933). By writing and lecturing widely Comrie brought awareness of machine computing techniques to the wider scientific community.

9.10 International Astronomical Union

Comrie was an active member of the International Astronomical Union and attended every meeting of the Union from 1928 onwards. Through these meetings Comrie met with many of the world's astronomers. As president of Commission 4 (Ephemerides) from 1932 to 1938, in succession to Ernest Brown, Comrie had a platform from which to influence astronomical table-making around the world. Apart from promoting the appropriate use of calculating machines, Comrie used his position to draw attention to the fact that while a great deal of international collaboration already occurred in sharing astronomical calculations still more could be done. His work in this area led to the publication of the *Apparent Places of Fundamental Stars* as an international resource (Figure 9.6).

FIGURE 9.6 L.J. Comrie (centre) with Johann Peters and K.H.W. Kruse at the 1930 International Astronomical Union meeting. (Reproduced from R.C. Archibald's *Mathematical Table Makers*, 1948, Scripta Mathematica No.3, courtesy of Yeshiva University.)

9.11 Non-Astronomical Table-Making

Outside of his duties at the NAO, Comrie was an active general mathematical table-maker. In 1931 he published a new edition of *Barlow's Tables of Squares, Cubes, Square Roots and Reciprocals*[37], and in collaboration with Louis Milne-Thomson (1891–1974), who taught Mathematics at the Royal Naval College (see Chapter 6), he published a set of four-figure logarithms designed for the non-specialist.[38] Another large table-making project to which Comrie was heavily committed during the early 1930s was conducted in collaboration with Johann Peters (1869–1941). Peters was one of the foremost mathematical table-makers of the twentieth century. He spent most of his working life at the Astronomisches Rechen-Institut in Berlin and was responsible for the production of many high-quality tables.[39] At the 1930 meeting of the Astronomisches Gesellschaft in Budapest Comrie and Peters discussed the possibility of undertaking a very large project to prepare a table of the natural trigonometric functions for every second of arc.[40] The ambitious plan had, because of lack of resources, to be reduced to producing two tables of the four functions sine, cosine, tangent, and cotangent; one to seven figures and one to eight figures. The division of labour on the project was simple. Peters provided the pivotal values and differences, and Comrie interpolated between them using the NAO Burroughs machine. When the project was finished in the mid-1930s, it proved difficult to get a publisher for the huge tables (approximately 10 million entries). In 1939 Peters succeeded in getting the eight-figure tables published in Germany[41], but the seven-figure tables remained in manuscript and were eventually deposited with the Royal Society in London.

9.12 British Association Mathematical Tables Committee

Many of the projects described above tied in well with Comrie's work at the NAO. While not strictly astronomical or navigational tables, they did have application in that field and Comrie collaborated with astronomers or colleagues from the Royal Naval College. However Comrie also contributed to and engaged with the wider mathematical and academic table-making community by becoming a member of the British Association for the Advancement of Science Mathematical Tables Committee. In 1871 the British Association for the Advancement of Science had realised that many generally useful mathematical tables were inaccessibly published in the periodical literature, and set up a committee to report on the available tables and make suggestions as to those tables that should be collected together and republished. When the Committee reported in 1873, it became obvious that simply reprinting tables would be too ambitious a project for the Committee and furthermore would not solve the problem of different intervals and layouts given in tables of the same function published by different people. Instead the Committee began to compute and, periodically, to publish tables under its own name.

Comrie was invited to serve on the Committee in 1928 and by 1929 had manoeuvred himself into the influential position of Secretary. The history of the Committee[42] clearly shows that Comrie's membership of the Committee corresponds to its most productive period. Under Comrie's leadership the Committee began to publish a series of high-quality mathematical tables. While Comrie did not undertake all of the calculations himself, his influence in the planning, computation, and typesetting of several volumes is very evident. At around the time that Comrie joined the Committee it received a bequest that Comrie encouraged the Committee to spend on purchasing several calculating machines and a National Accounting machine. As described above Comrie housed the National Machine at the NAO where it was used for Committee work. Comrie was influential in bringing two young table-makers, J.C.P. Miller and Donald Sadler, who both used machines installed at the NAO for British Association Mathematical Tables Committee work. In 1930 both Miller and Sadler had applied for the post of Deputy Superintendent at the NAO; the Civil Service turned down Miller on health grounds, but Sadler was accepted. On Sadler's first day in post he found himself working with Miller on a Mathematical Tables Committee table using an NAO Brunsviga.[43]

9.13 Consultancy Work

Because of the high profile Comrie generated for himself many people came to him for advice and help. Documents in the Royal Greenwich Observatory archives[44] clearly demonstrate that both calculating machine manufacturers and sales agents consulted him professionally, as did academics and government departments wanting to set up their own computing facilities.[45] For example in 1934 the University of Liverpool wanted to set up a library, and Comrie not only advised but also donated offprints of his papers. He did likewise for the University of Aberdeen. Douglas Hartree (1897–1958), a mathematical physicist and expert on numerical computation, also took advice from Comrie on what make and model of calculating machine would be most suited to the needs of the mathematics and physics departments at Manchester University. Through Hartree, Comrie also advised the X-ray crystallographer Lawrence Bragg on how he could apply Hollerith punched card machines to his work. The University of Calcutta wrote to him asking for advice on what calculating machines to buy and Comrie gave them his opinion. The Ballistics Research Laboratory too consulted Comrie before purchasing a National Accounting Machine, and he advised the Ordnance Survey. Comrie continued his relationship with Karl Pearson, and later his son Egon, at the Biometrics Laboratory at University College, London, and contributed to the Laboratory's *Tracts for Computers* series.[46]

Comrie also gave practical assistance to those who asked for his help – although his inability to suffer fools gladly made this a nerve-racking experience for those who were brave enough to approach him. One of those who did was Gordon Hey, a young mathematician working at Cambridge on the optimum sample sizes of sugar beet for an agricultural investigation.[47] Hey and his colleague Hudson approached Comrie following a presentation he had given to the Royal Statistical Society. Comrie invited them to the NAO to discuss the problem and advised them that Hollerith punched card machines would be the most efficient way to undertake the statistical analysis they needed. Through Comrie's links with the British Tabulating Machine Company, the UK manufacturers of Hollerith machines, Hey and Hudson not only obtained access to the machines they needed but also had them specially modified to make them more suitable for the work.

As Comrie and the NAO's reputation as a centre of computational excellence grew, so did the number, size, and importance of requests that Comrie was receiving. In 1934 Comrie undertook a series of triangulation calculations for the Geographical Section of the War Office. Although performed outside normal NAO working hours, Comrie used NAO staff (paid overtime directly by Comrie) and NAO calculating machines. The Admiralty discovered the arrangement and forbade Comrie from carrying out any further work of this type without prior authorisation.[48]

However in 1935 Comrie received a request from the War Office to undertake sound-ranging calculations. As part of its preparations for war, the War Office was trying to improve the accuracy with which it could pinpoint the location of enemy artillery fire – work they perceived to be of national importance. Given his previous rebuke from the Admiralty, Comrie officially requested that NAO facilities be used to assist the War Office, but permission was refused on the grounds that the Admiralty requirement for the NAO was of higher priority.[49] Although the NAO work was almost complete, Comrie had been trying to argue for more staff and had led the Admiralty to believe that NAO work was behind schedule – a tactic which backfired on him in this instance. The War Office contracted the work out to a company called Calculating and Statistical Service, who in turn subcontracted the work privately to Comrie.

9.14 The End of an Era

By the mid-1930s Comrie was using NAO staff to help with some of the outside work, paid and unpaid, that he was undertaking. Conversely Comrie was privately employing and privately paying computing staff who sometimes worked in the NAO using NAO calculating machines. This mix of public and private work was a complete anathema in the Civil Service of the mid-1930s. While Comrie had a high international reputation as a world class mathematical table-maker, relations between Comrie and his Admiralty employers were often strained. Not only was Comrie always asking for more staff, but he also repeatedly asked for larger accommodation at either the Royal Naval College or elsewhere. During the mid-1930s, the Admiralty suggested that the NAO move to Trafalgar Quarters situated on Park Row in Greenwich (and previously lodgings for officers of Greenwich Hospital) or Devonport House on King William Street. Comrie thought neither suitable. In 1936 the Royal Naval College offered the NAO extra rooms within the College. The NAO did not leave the College until 1939.[50]

Having forbidden Comrie from taking on the War Office sound-ranging work, an Admiralty Board of Enquiry paid an unannounced visit to the NAO in August 1936 to discover that he was indeed using NAO staff for the work, albeit not directly for the War Office.[51] Comrie was immediately suspended from duty and eventually dismissed. The irony is that during the war the NAO was officially required to take on the sound-ranging and other official computing work for various parts of the armed services.

Comrie's reputation and level of consulting work led him almost immediately to set up the Scientific Computing Service as a commercial company specialising in table-making and other computing work. Through the Scientific Computing Service Comrie continued his consultancy table-making work – but this time on a commercial footing. His work for the armed services throughout the Second World War was very highly regarded.[52] In the post-war period Comrie is best remembered for his revision of Chambers *Mathematical Tables* and his work on identifying and publishing errors in tables,[53] but he also applied punched card machines to the Fourier synthesis work required by X-ray crystallographers. Even after his dismissal from the NAO Comrie continued to live on Maze Hill next to Greenwich Park. Here he entertained travelling New Zealanders and mathematicians and scientists with unfailing hospitality until his death in December 1950.

10

Artful Measures: Mathematical Instruments at the National Maritime Museum

Richard Dunn

CONTENTS

10.1 Introduction

On opening to the public in 1937, the National Maritime Museum (NMM) could not have found a more natural home than Greenwich, a place so bound up with Britain's scientific and maritime past.[1] The NMM collections have grown considerably since then, but have always included rich holdings of scientific instruments. Some of these relate to Greenwich's specific role in the development of mathematics, while others are witness to the more general evolution of the mathematical sciences broadly construed. This chapter looks at some of the mathematical instruments now in the Museum and the stories behind them.

10.2 What Is a Mathematical Instrument?

The term 'mathematical instrument' began to be established from the sixteenth century as a specialist scientific instrument-making trade developed and expanded, initially in London. By the eighteenth century, English makers dominated world markets, and three categories of instrument had come to be recognised: mathematical, philosophical, and optical.[2]

Although distinctions between the three categories were never absolute, some general observations are possible. Mathematical instruments were those used for number-based measurements and generally had some sort of divided scale. Philosophical instruments, such as air pumps, were used to investigate, but not necessarily to quantify, natural phenomena, while optical instruments made use of lenses, mirrors, and prisms. Nevertheless, there was considerable overlap between the three, such as the use of telescopes and graduated scales on instruments like theodolites, as well as crossover in the sorts of products makers and retailers offered for sale.[3]

10.3   Ships and Stars: The National Maritime Museum and Its Collections

The NMM's origins can in fact be traced to the nineteenth century, with the opening in 1824 of a National Gallery of Naval Art in the Painted Hall of what became the Royal Naval College, Greenwich, which also housed a Naval Museum from 1872–1873. It was only in 1934, however, that the Museum was created under the National Maritime Museum Act. Although attempts to assemble a collection had already begun in the previous decade, once the Museum was formally established, the contents of the Naval Museum were transferred to its care along with other official material, and the Greenwich Hospital Collection from the Painted Hall was added as a long-term loan. Finally, on 27 April 1937, King George VI officially opened the NMM, and visitors were admitted two days later. Since then, the Museum collections have grown and now comprise, by some counts, nearly 2.5 million items, including internationally important holdings of maritime art, cartography, manuscripts, ship models and plans, timekeepers, and scientific, astronomical, and navigational instruments – a collection 'unsurpassed in Britain and scarcely equalled in Europe'.[4]

Mathematical instruments, in particular those relating to navigation and cartography, were of interest right from the start. Many of the early acquisitions in this area, as in others, were funded by Sir James Caird, notably the purchase of George Gabb's collection of over 1,300 scientific instruments. Alongside pre-assembled collections like Gabb's, the NMM has also acquired groups of objects that reflect professional lives, such as that of the Reverend George Fisher (1794–1874). Having served as astronomer on the Admiralty Arctic expedition of 1818 and William Parry's North-West Passage expedition of 1821–1823, Fisher later became headmaster of the Greenwich Hospital Schools. His collection readily fitted the Museum's collecting priorities.[5]

The NMM's interest in mathematical instruments broadened further in the 1950s, when the former Royal Observatory buildings in Greenwich Park were transferred to its care as the working astronomers completed their move to Herstmonceux Castle in Sussex.[6] This led to a more explicit interest in astronomy, with the Royal Observatory also becoming the focus for the Museum's scientific displays (Figure 10.1), although the maritime galleries also highlight some aspects of the role of mathematics and mathematical instruments in maritime history.

Today, the NMM's collection of scientific and mathematical instruments has strengths in areas that include

- Astronomy, from large observatory instruments to hand-held instruments such as quadrants and astrolabes[7]
- Navigation, including compasses, and sextants and related devices[8]
- Chronometers, clocks, and watches[9]
- Sundials and other non-mechanical timekeeping devices[10]
- Cartography

FIGURE 10.1 Sundials displayed in the Royal Observatory, Greenwich. (National Maritime Museum Neg. F5118-13 © National Maritime Museum, Greenwich, London.)

- Magnetic and meteorological research
- Surveying
- General mathematics, including computing instruments and mathematical and drawing instrument sets.

Together, these objects illustrate not only the activities of mathematicians working in Greenwich, but also more general narratives, often related to the interests of those working in and around this part of London. The following selection illustrates some of these stories and the range of instruments in the collection.

10.4 At the Cutting Edge: The Royal Observatory and Technical Innovation

King Charles II's founding warrant of 1675 defined the purpose of the new Royal Observatory at Greenwich as 'the finding out of the longitude of places for perfecting navigation and astronomy'.[11] In pursuing this purpose, the Observatory and its staff assumed an influential role in the development of precision observing instruments, many of which can still be seen there (Figures 10.2–10.4).

From the start, the Observatory's main role was to create accurate star charts as the basis for tables that would allow navigators to determine their position at sea. John Flamsteed (1646–1719), first Astronomer Royal, began this work as soon as he took up residence in 1676, with his star catalogue published posthumously in 1725.[12] Charged with continuing this work, his successor, Edmond Halley (1656–1742), was given £500 to re-equip the Observatory – Flamsteed's widow having unfortunately removed the original instruments.

One of the instruments Halley commissioned was a large mural quadrant (Figure 10.2) from George Graham (about 1673–1751). Graham was by this time a leading London instrument-maker, whom the French mathematician Pierre-Louis Maupertuis would later hail as an 'ingenious Artist' due to the accuracy of his

FIGURE 10.2 Mural quadrant, by George Graham, 1725. (National Maritime Museum Neg. L2157-1 © National Maritime Museum, Greenwich, London.)

instruments.[13] While Graham sub-contracted most of the production of the Greenwich quadrant, he personally carried out the crucial work of dividing the scales (a task that was done at the Royal Observatory). There were two of these: an inner scale divided into conventional degrees and an outer scale divided into 96 parts (with further sub-divisions) by the more certain process of repeated bisection. A table for converting these parts into degrees, minutes, and seconds allowed the two scales to be cross-checked. The mural quadrant was installed in September 1725, with the first observations made on 27 October. It immediately proved its accuracy, which was partly due to the stability of its heavy iron frame,[14] but mostly to the precision of Graham's scales. Indeed, even sixty years later its performance inspired William Ludlam to note that Graham had 'carried the art of constructing and graduating... to such perfection, that from this time we may date a new era in Astronomy'.[15]

With the quadrant as his principal observing instrument (and remaining in use at the Observatory until 1812), Halley was able to carry out a long programme of observations to support the lunar-distance method for determining the longitude (see below). The successful demonstration of the instrument's accuracy also meant that it became the prototype for quadrants commissioned for observatories throughout

THE TRANSIT-CIRCLE AT THE ROYAL OBSERVATORY, GREENWICH.

FIGURE 10.3 Airy's transit circle, from E. Dunkin, *The Midnight Sky* (London, 1891). The astronomer on the right is reading the scale. (National Maritime Museum Neg. F5926 © National Maritime Museum, Greenwich, London.)

Europe. Indeed, the expanding export market enjoyed by English instrument-makers in the late-eighteenth century owed much to the Greenwich quadrant and the international reputation that George Graham, and thus the London trade in general, earned from it.[16]

The ongoing quest for accuracy in the instruments commissioned kept the Royal Observatory at the forefront of precision astronomy for the next two centuries. This arguably reached its zenith in 1850 with the installation of the great transit circle (Figure 10.3), the instrument that came to define the world's Prime Meridian.

Following his 1835 appointment as seventh Astronomer Royal, George Biddell Airy (1801–1892) began to expand the role of the Observatory beyond just supplying data for the Navy and the British Empire, aiming to turn it into a leading scientific research

FIGURE 10.4   Marine sextant, by John Bird, London, about 1758. (National Maritime Museum Neg. F4928-1 © National Maritime Museum, Greenwich, London.)

institute, with initiatives that included the creation of a department for magnetic and meteorological data. Nonetheless, Airy acknowledged that the Observatory's main purpose was still the production of accurate astronomical tables and, like his predecessors, wished to have the best equipment to achieve this. This desire was evident in the new transit circle, which was to be more optically powerful and more accurate than its predecessors.

Having secured a large, high-quality lens, Airy personally designed the instrument to house it, sub-contracting the manufacture of the cast-iron superstructure to Ransome and May of Ipswich and of the optical and instrumental parts, which were manufactured to the highest standards of precision, to William Simms of London. At the heart of the new instrument was an observing telescope that rotated in the plane of the meridian (a north-south line). Readings were taken from a 6 ft circle fixed to one side, with its degree scale read with six microscopes. There were normally at least two observers, one looking through the telescope, the other reading the scale through the microscopes (see Figure 10.3), with each read off in turn and their average taken to give observations accurate to one thousandth of a minute of arc.

The instrument could be used in a number of ways. The observers could fix a star's position by recording its vertical angle and the exact time on a regulator (a highly accurate clock) as the star passed across the meridian. The transit circle could also be used to set and regulate the Observatory's standard clocks (from which Greenwich time was distributed) by observing a small, well-recorded group of 'clock stars' as they passed across the instrument's meridian.[17]

With the instrument completed in 1850, Airy intended to make the first observation on 1 January 1851, the first day of the new half-century, but was frustrated until three days later by poor weather – a common British problem. For the next 103 years, however, the instrument was used almost continuously to make over 1,000,000 observations and determinations. Many aspects of its revolutionary design were also adopted for similar instruments in observatories worldwide. Its main claim to fame, however, is that at the International Meridian Conference in Washington D.C. in 1884, it was recommended that the world's Prime Meridian would be defined as the meridian passing through the centre of the instrument. It is this line that visitors from all over the world come to see.[18]

As well as being a client for these large instruments, the Observatory's position at the cutting edge of astronomy and its involvement in navigational matters gave it a role in the development of other types of mathematical instrument. One of these had a major part to play in solving the longitude problem that had taxed mathematicians and navigators for centuries.

Longitude is the east-west co-ordinate of a geographical position. Since this is a measurement in the direction of the Earth's rotation, the longitude difference between two places can be thought of as the difference between their local times (the time as indicated by the Sun's position, local noon being the point when the Sun appears to be at its highest in the sky). Thus, many methods for determining longitude at sea focussed on trying to ascertain simultaneously the ship's local time (from the Sun) with that at some reference location, such as a port or astronomical observatory. By the eighteenth century, the anticipated solution seemed to rely on the successful accomplishment of one of a handful of methods, including longitude by lunar distance and by mechanical timekeeper. Work on these methods (and others) received a huge impetus from the 1714 Longitude Act and its rewards of up to £20,000 for a successful solution. The Act also created a group of Commissioners (later known as the Board of Longitude), who included the Astronomer Royal and were to encourage the development of promising ideas and oversee the granting of rewards.[19]

In the end, the lunar-distance and timekeeping methods were both shown to be practicable within a few years of each other in the 1760s, and Greenwich had a role in each story. The successful creation of an accurate marine timekeeper by John Harrison has been well documented,[20] but just as significant was the work done on the lunar-distance method that culminated in the development of the marine sextant (Figure 10.4) and of astronomical tables to be used with it.

The lunar-distance method uses the fact that the Moon travels by approximately its own diameter in one hour against the background of the fixed stars. In principle, this allows an observer who knows its future positions to calculate the time at some known reference point. By determining the local time where they are (e.g. by observing the Sun), the observer can work out the difference between the two times and hence their longitude.[21] Although the principle was straightforward, two practical problems faced anyone promoting this method. Firstly, the complexity of the Moon's motions meant that no accurate tables of its movements, actual or predicted, were yet available. To this end, Edmond Halley, second Astronomer Royal, began a long programme of lunar observations with the large mural quadrant by Graham, and his successors at Greenwich and other astronomers in Europe carried on the work after his death.

Yet even if this work were to make accurate tables available, the second problem was that there was no instrument with which to make sufficiently accurate observations from a moving ship. This unpromising situation began to change in the 1730s with the invention of the marine octant, developed simultaneously by John Hadley (1682–1744) in England and Thomas Godfrey (1704–1749) in America. The innovation in these designs was to measure the height of a star by using mirrors to bring a reflected image of the object alongside an image of the horizon.[22] As Alan Stimson has observed, the octant 'enabled latitude to be observed with unprecedented accuracy at sea and also the altitude of the Sun with sufficient accuracy for the local time to be determined on board ship – both prerequisites to determining longitude'.[23]

Hadley's octant was successfully tested in 1732 in trials that involved a young astronomer named James Bradley (1693–1762).[24] Succeeding Halley as Astronomer Royal ten years later, Bradley began an important relationship with another London instrument-maker, John Bird (1709–1776). This association would have practical implications for the Observatory and for the development of marine navigation. In 1749–1750, Bird constructed a transit instrument and a mural quadrant of 8 ft radius (following Graham's design for the first mural quadrant) for the Observatory,[25] and his ongoing collaboration with Bradley meant that the instruments and techniques in use at Greenwich began to set new standards in terms of accuracy and precision.[26]

Like George Graham, Bird supplied instruments to observatories throughout Europe. These included a mural quadrant (again of Graham's design) for the observatory at the University of Göttingen, where Tobias Mayer (1723–1762) created accurate astronomical tables aimed at perfecting the lunar-distance method. Mayer also devised a new hand-held instrument, a repeating circle, similar in function to an octant but with a complete circular scale. After some persuasion by his colleagues, Mayer submitted these and a method for their deployment at sea to the Commissioners of Longitude, who instructed Bradley 'to cause Three Instruments to be made ... for the taking Observations necessary to be made on Ship Board in the intended trials of Mr Meyer's [sic.] Method for finding the Longitude of a ship at Sea'.[27] Unsurprisingly, Bradley had Bird make the instruments, which included a reflecting circle, and they were tested with Mayer's tables by Captain John Campbell (1720–1790) on the *Chatham* in 1757. Campbell, later described as 'almost unrivalled as a seaman, and among seamen, perhaps, wholly so, as an astronomer and navigator', seems to have been a good choice.[28] Finding the circle rather cumbersome to handle during the sea trials, he asked Bird to make a more manageable instrument with a scale that measured to only 120 degrees. Campbell was then able to test this, the first marine sextant, during the Royal Navy's blockades of the Atlantic ports of France in 1758–1759, and before long Bird was making further copies, some of which survive (Figure 10.4).[29]

Mayer's tables and the sextant helped revolutionise navigational practice and were the keys to solving the lunar-distance method. This found its champion in the fifth Astronomer Royal, Nevil Maskelyne (1732–1811), who also oversaw the creation and production of the *Nautical Almanac*, astronomical tables produced by the Royal Observatory to aid sextant- and chronometer-based navigation at sea.[30] It is most appropriate, therefore, that one of the first marine sextants is on display in the Royal Observatory.

10.5 Everyday Instruments for Practical Purposes

Not all the mathematical instruments held at the NMM are cutting-edge technical innovations. Many are simply tools for practical daily situations. The horary quadrant in Figure 10.5 is one of these, but it also embodies one of the most significant changes in eighteenth-century English life – the adoption of the Gregorian calendar.

A quadrant is a hand-held instrument that can be used to tell the time, measure heights and distances, and carry out astronomical and mathematical calculations. While these instruments had been known for centuries, it was during the sixteenth century that many new forms began to appear, including versions like this one that would give the time in equal hours.[31] The one drawback of these newer quadrants was that they could only be used for a single latitude – in this example for 51°32′ (Figure 10.6), which was assumed to be the latitude of London.[32]

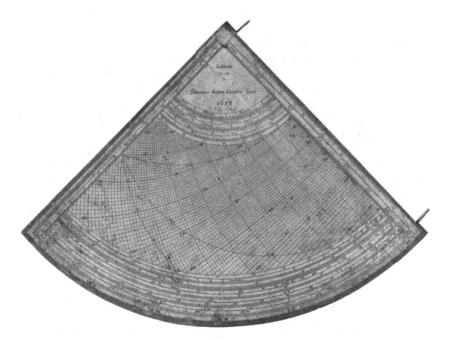

FIGURE 10.5 Horary quadrant, after Henry Sutton, about 1750. (National Maritime Museum Neg. F5702-1 © National Maritime Museum, Greenwich, London.)

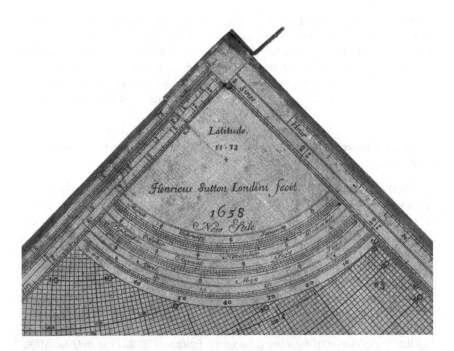

FIGURE 10.6 Detail of the horary quadrant in Figure 10.5. Residual marks on the declination scales show that the printing plate was re-engraved to alter them to the 'New Stile' calendar. (National Maritime Museum Neg. F5702-3 © National Maritime Museum, Greenwich, London.)

It was either Henry Sutton (*c.*1624–1665) or Thomas Harvie who designed this form of quadrant, although it now carries Sutton's name.[33] According to the title of Sutton's own description, its many uses included 'the easie resolving of all astronomical, geometrical, and gnomonical problems, ... working of proportions, and... finding the hour universally'.[34] Although it was similar to other quadrants, Sutton's successful promotion ensured that by the end of the seventeenth century, it was one of the most commonly used astronomical calculating devices in England. Indeed, as late as 1758 Edmund Stone remained adamant that 'Mr Sutton's Quadrants, made above one hundred Years ago, are the finest divided Instruments in the World; and the Regularity and Exactness of the vast Number of Circles drawn upon them is highly delightful to behold'.[35]

What is significant about this particular example is the appearance of the words 'New Stile' under the date of 1658 (Figure 10.6). This refers to the competing calendar systems operating in Europe in the seventeenth and eighteenth centuries. Much of mainland Europe was following the Gregorian (new style) calendar by the late seventeenth century. This had originally been proposed by Pope Gregory XIII in 1582 to correct errors arising in the Julian calendar, and, in the decades following Gregory's papal decree, most Catholic countries adopted the new calendar. England, like a number of Protestant countries, persisted with the Julian.

Britain proved particularly resistant to the Gregorian calendar. Indeed, it was not until the mid-eighteenth century that change finally took place, with the Fourth Earl of Chesterfield (1694–1773) as its champion. Chesterfield's main concern, however, was not the inaccuracy of the Julian calendar, but the eleven-day difference in dates that had arisen between Britain and Europe. While serving as an ambassador in France, he had found it particularly confusing to have to convert between the two calendars. This was specifically alluded to in the new Act, which stated that

> it will be of general convenience to merchants and other persons correspond-
> ing with other nations and countries, and tend to prevent mistakes and dis-
> putes in or concerning the dates of letters and accounts, if the like correction
> be received and established in his Majesty's dominions.[36]

Chesterfield was no astronomer and freely admitted it. He later confessed that to push the reform through, 'I was obliged to talk some astronomical jargon, of which I did not understand one word, but got it by heart, and spoke it by rote from a master'.[37] It was his more mathematically literate acquaintances who fleshed out the details. These included Lord Macclesfield, a leading astronomer (and later president of the Royal Society) and friend of James Bradley. Completing the network was the Prime Minister, Henry Pelham, whose wife was Ranger of Greenwich Park. Together, this small group created and put forward the bill for calendar change, which was drawn up in 1749–1750.

As Astronomer Royal, Bradley's role was to ensure that the astronomical and mathematical details were sound. In particular, he had to show that the dating of Easter could be brought into line with that of the rest of Europe, yet avoid an osten-sibly Catholic method of calculation. Thus, his revised tables, which notionally used the Anglican method, ensured that the dates matched those calculated using the Gregorian system. The resulting 'Act for Regulating the Commencement of the Year; and for Correcting the Calendar Now in Use' was passed in 1751. As a result, 1752 started on 1 January (rather than 25 March, the start of the English legal year until then) and Wednesday 2 September 1752 was followed by Thursday 14 September in order to drop the eleven days required to bring Britain in line with Europe. From then on, this country was finally using the 'New Stile' calendar.[38]

This brings us back to the quadrant, since one of the scales it carries is a date scale, used to 'set' the instrument for the appropriate declination of the Sun, which varies throughout the year (Figure 10.6). When the quadrant was issued in 1658, this scale accorded with the Julian calendar.[39] To use the quadrant with the 'new style' dates of the Gregorian calendar, the scale had to be altered by eleven days, which is what hap-pened when the quadrant was re-issued in the mid-eighteenth century.[40]

There is a further Greenwich connection, albeit a somewhat spurious one, since a number of myths surround the 1752 calendar reform. The most famous is that the 'loss' of eleven days was the cause of rioting throughout England.[41] In similar vein, James Bradley's first biographer reported that during the Astronomer Royal's final illness, 'many of the common people attributed his sufferings to a judgement from heaven for his having been instrumental in what they considered to be so impious an undertaking'.[42]

Looking at other instruments from the NMM collection, one that demonstrates a practical solution to mathematical problems is known as Napier's bones or Napier's rods (Figure 10.7). Its invention arose from the creation of logarithms by John Napier (1550–1617), the use of which radically improved the speed, accuracy, and power with which mathematicians could perform computations. Hoping to eliminate 'the

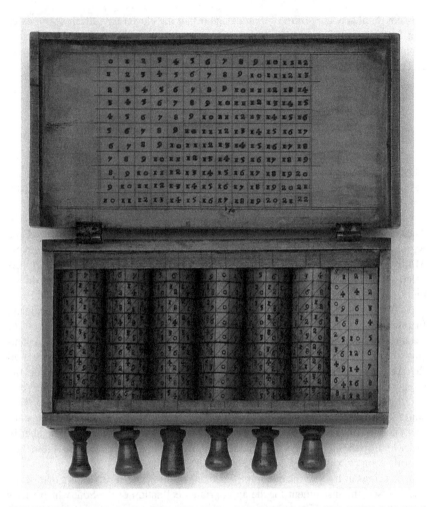

FIGURE 10.7 Napier's bones, about 1679. (National Maritime Museum Neg. D7195 © National Maritime Museum, Greenwich, London.)

tedium and difficulty of calculation', which he believed deterred many from studying and doing mathematics, Napier went on to publish *Rabdologiae, seu Numerationis per Virgulas* (1617). This described three instruments intended to make computation easier. The simplest was a set of rods, on each face of which was the multiplication table, from one to nine, of a digit. The different rods could then be used in combinations to perform mathematical functions such as multiplication and division.[43]

The new mathematical aid soon came to be known as 'Napier's bones' and gained in popularity. Samuel Pepys, for example, recorded in his diary in 1667 that Jonas Moore had instructed him in 'the mighty use of Napier's Bones'.[44] Pepys, then aged 29, had recently become Clerk of the Acts of the Navy and was employing Moore (1617–1679) to teach him arithmetic.[45] Moore had been a mathematical teacher in London since the 1640s and was appointed Surveyor-General of the Ordnance two years after Pepys's diary entry. As a patron of the 24-year-old John Flamsteed, he was also one of the group who persuaded Charles II to build the Observatory in Greenwich and to appoint Flamsteed as first Astronomer Royal.[46]

Napier's bones also found their way into literature. In the second book of Samuel Butler's mock-epic poem *Hudibras* (first published in 1663), the devices used by the knavish astrologer Sidrophel include 'A *Moon-Dial*, with *Napiers* bones, And sev'ral *Constellation*-stones'.[47] Back in the real world, later improvements to Napier's bones included making them as a box containing ten cylinders, each divided into ten strips, allowing the numbers to be 'dialled' rather than selected from loose rods. This design, which is the type illustrated in Figure 10.7, was being made in London by the late seventeenth century. In 1710, for example, Zacharias von Uffenbach wrote that he purchased 'two very convenient kinds of Bacillus Nepperianis' from an instrument-maker in Westminster Hall, one of them consisting of 'round rods which are turned in front by means of little knobs'.[48]

Sadly, Napier's rods had a more limited life than the Observatory. Over the course of the eighteenth century, such aids became increasingly consigned to the classroom as written arithmetic became the favoured way of demonstrating mathematical ability. By the early nineteenth century they were rarely used.[49]

One area of practical mathematics in which the NMM has a particular interest is navigation, and the Museum's navigational holdings are among the best in the world. While they include examples of innovation, most are instruments that were routinely used on board ships every day. A group that typifies this was recovered from the *Stirling Castle*, one of several ships wrecked on the Goodwin Sands, off the Kent coast, during the Great Storm of 1703.

The Great Storm began on Friday 19 November 1703 and lasted 13 days. Described by eyewitnesses as 'the greatest, longest and most severe storm that ever the world saw',[50] its effects were felt across Europe, but were at their fiercest in south-east England, where the winds ripped off roofs, blew fish from the rivers, flattened trees and steeples, lifted a cow into a tree, and killed many people. In London, ships were blown along the Thames and ground each other to pieces. The effects were even more devastating at sea, with over a hundred vessels wrecked or scattered and the Eddystone Lighthouse swept away with its builder, Henry Winstanley (about 1644–1703).[51]

Midway through the storm, there was a period of relative calm, and on 25 November, a fleet of ships under Sir Cloudesley Shovell (about 1650–1707) entered the Downs and anchored, thinking the worst was over. Yet by the morning of 27 November, the storm was picking up again. Stephen Martin, flag captain on the *Prince George*, recorded the terrible sight:

> At three in the morning the *Restoration*, dragging her anchors, came down upon us, for a half hour the ships beat against each other, the longest half hour we ever knew, but at last the invisible hand of Providence relieved us, the *Restoration* drove away into the night and was lost with every living creature aboard. Next day we saw 12 sail ashore upon the Goodwin, Burnt Head and Brake Sand, amongst them Admiral Beaumont in the *Mary*, the *Stirling Castle*, *Northumberland* and *Restoration*, who were all to pieces by ten o'clock, with all perished, except one of the *Mary* and seventy from the *Stirling Castle*. It was a melancholy prospect to see two and three thousand perish without any possibility of helping any of them.[52]

Amazingly, when archaeologists rediscovered the *Stirling Castle* in 1979, almost everything except iron and human flesh was still preserved. She therefore provides first-class evidence of the equipment carried by an English ship of the line around 1700.[53] Among the wealth of items recovered – bottles, lamps, candle snuffers, shoe buckles, pewter, pieces of fabric, and much else – divers found a seaman's chest crushed under a gun barrel on the lowest deck. Inside were navigational instruments, including dividers, sand-glasses, the remains of at least one cross-staff (Figure 10.8), and a Gunter rule (Figure 10.10).[54] These last two are particularly interesting because both instruments were widely used yet rarely survive because they were fairly cheap and made of wood rather than metal.

The cross-staff was invented in the fourteenth century and used in navigation by the early sixteenth century. Its main use at sea was to measure the angle of the sun or a star (usually the Pole Star) above the horizon in order to work out the ship's latitude. This was done by sliding a cross-piece (vane) along the staff until one of the vane's ends lined up with the horizon and the other with the star (Figure 10.9). The observed angle could then be read from a degree scale inscribed on the staff. There was usually one scale on each face, with four vanes of different lengths, one for use with each scale. These allowed the more accurate measurement of different angles.[55] Cross-staves were made in great numbers until the mid-eighteenth century, when the octant and sextant began to become more common. Nonetheless, they were still being produced in the Netherlands until the early nineteenth century.[56]

The Gunter rule or Gunter scale (Figure 10.10) was designed to help navigators with the calculations for plotting their course and position. Conceptually, it can be seen as the link between Napier's logarithms and the slide rule. The principles were described by Edmund Gunter (1581–1626) in *The Description and Use of the Sector, the Crosse-Staffe and other Instruments* (1624), which suggested that logarithmic scales would make calculations quicker and easier for practical situations such as calculating a ship's position. It is not clear, however, when the Gunter rule

FIGURE 10.8   Parts of a cross-staff, about 1700, from the wreck of the *Stirling Castle*. (National Maritime Museum Neg. F4931-1 © National Maritime Museum, Greenwich, London.)

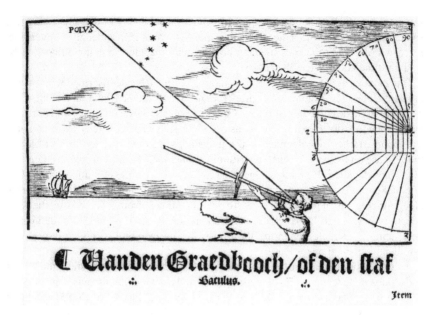

FIGURE 10.9   Using a cross-staff, from Cornelis Anthonisz, *Onderwijsinge vander zee om stuer-manschap te leeren* (Amsterdam, 1558). (National Maritime Museum Neg. A9060 © National Maritime Museum, Greenwich, London.)

FIGURE 10.10   Gunter rule (detail), by John Arnold, 1700, from the wreck of the *Stirling Castle*. (National Maritime Museum Neg. F5700-003 © National Maritime Museum, Greenwich, London.)

was introduced, since it does not appear in Gunter's published works, suggesting that someone else applied the scales to a fixed rule. Yet whoever invented the Gunter rule, its ease of use and robustness made it popular with mariners, and it continued to be sold until the late nineteenth century.

Gunter rules are commonly 2 ft (61 cm) long, with each side inscribed with twenty or more scales. The example in Figure 10.10 has scales on the front that include a linear scale in inches; a diagonal scale to set the dividers to exact lengths; and scales of leagues, chords (for measuring or constructing angles), rhumbs (relating to compass rose points, rather than degrees), sines, tangents, semi-tangents, and miles of longitude. Those on the back include logarithmic scales, such as sines of rhumbs, tangents of rhumbs, logsines, logversines, logtangents, as well as a line of numbers and a line of equal parts. Each scale allows the relevant function to be calculated by measuring along the rule with a pair of dividers. To prevent damage from repeated use, the rules generally have small brass pins at the most frequently used points. Although cheap and seemingly simple, therefore, these instruments facilitate surprisingly sophisticated mathematical operations.[57]

As well as individual instruments like the Gunter rule and cross-staff, the NMM collections include many sets of mathematical and drawing instruments. One of these (Figure 10.11) was awarded in 1901 to Walter Theodore Bagot (born 1885) as a prize in his final exams in Charts and Instruments from HMS *Britannia*. The set includes a rule and protractor, a parallel rule, and several dividers, compasses, and pens – typical instruments for working with sea charts.

FIGURE 10.11   Mathematical instrument set, about 1901. (National Maritime Museum Neg. F5703-001 © National Maritime Museum, Greenwich, London.)

The same set also tells us something about the teaching of navigation and seaman-ship in the period. *Britannia* was one of the Royal Navy's training ships and had been moored at Dartmouth since 1869 for cadet training.[58] By the time Bagot was there, about 65 cadets, aged about 15, were coming on board every four months to be trained for four terms. Their classes included mathematics, navigation, nautical astronomy, instruments, charts, steam machinery, French, drawing, and seamanship, with exams at the end of the second and fourth terms. Those who passed became Naval Cadets or Midshipmen, depending on their conduct and final results, with prizes like Bagot's awarded after the final exams.[59]

Britannia was one of a number of establishments where boys and men could train for a career in the Royal Navy. In Greenwich, the Royal Hospital School, which from 1821 to 1933 occupied the Queen's House and other buildings that are now part of the NMM, also taught navigation and seamanship. Its history had started much ear-lier than *Britannia*'s, with 'orphans of the sea' taken into the Hospital from 1715 and taught navigation nearby at an Academy set up by Thomas Weston, who until 1706 had been an assistant to John Flamsteed at the Royal Observatory. By 1841, boys between the ages of five and twelve were entering the School before joining the Royal Navy or the Royal Marines.[60] At a higher level, the Royal Naval College occupied the buildings opposite the NMM (now used by the University of Greenwich) from 1873 to the 1990s and provided 'education of Naval Officers of all ranks above that of midshipman in all branches of theoretical and scientific study bearing upon their profession'.[61]

FIGURE 10.12 Station pointer, by Potter, about 1875. (National Maritime Museum Neg. F5701-001 © National Maritime Museum, Greenwich, London.)

A final example (Figure 10.12) illustrates another set of stories that relate in part to Greenwich. The instrument is a station pointer, which consists of a graduated circle with one fixed and two moveable arms that rotate around the centre of the circle. This arrangement allows an observer's location to be plotted on a chart by measuring the relative angles of three known objects (usually land features) with a sextant, and setting the moveable arms to the observed angles. Placing the instrument on the chart and aligning the arms with the observed objects, the centre of the circle indicates the observer's position. The instrument's principal use was for hydrographic surveying, which was the context for which it was devised in the 1770s, probably with the help of the aforementioned John Campbell. Thus, station pointers were mainly to be found on naval vessels, where they formed the backbone of British hydrographic achievements in the nineteenth century.[62]

This example has had a varied history, as its inscriptions show. The maker's name on it is 'Potter Poultry London', referring to John Potter who worked in the City of London and became instrument-maker to the Admiralty.[63] On the central arm is the British government mark (a broad arrow) beside the inscription '2 Polar Exp. 1875', testifying to its use on the Royal Navy's Polar expedition of 1875–1876 under the command of George Strong Nares (baptised 1831, died 1915).

Nares first entered the navy in 1845 from the Royal Naval School in New Cross[64] and in 1858 joined the staff of the training ship *Illustrious*, the predecessor of the first *Britannia* at Dartmouth. During his time there, Nares wrote *The Naval Cadet's Guide* (1860), later re-titled *Seamanship*, which came to be regarded as the best manual of its type. In 1865, however, he resumed his seagoing career, undertaking survey work on the coast of Australia and in the Mediterranean. This paved the way for his appointment in 1872 as captain of HMS *Challenger* on what is often described as the first voyage to undertake systematic oceanographic research, and then as leader of the 1875 Arctic expedition.[65]

The primary aim of the 1875 expedition was to reach the North Pole. In addition, Nares's sailing orders made it clear that 'the object of the expedition is for the advancement of science and natural knowledge', though adding that the 'requirements of hydrography and geography will be provided for if the prominent features and general outline of the shores are sketched in as faithfully as circumstances will admit; and to ensure recognition by future explorers'.[66] Surveying was only a secondary purpose, therefore, although the expedition successfully charted 250 miles of coastline with the help of this station pointer, despite being cut short following an outbreak of scurvy.

The same instrument has a further history. As an additional inscription indicates, it was used on the river Thames, first by the Thames Conservancy and then by the Port of London Authority (PLA).[67] Today, the PLA is still responsible for vessel traffic management, maintaining shipping channels and moorings, and positioning buoys and markers. As part of this work, staff carry out periodic surveys of the river, including the Greenwich waters, using equipment that until the 1970s included this station pointer. This very practical instrument has seen over a century of active service.

10.6   Conclusion

These examples have been highlighted to indicate the range of mathematical instruments in the collections of the National Maritime Museum, as well as to highlight how many of these instruments relate to the work that has been undertaken in and around Greenwich in the past 350 years. They are only a small taster, however, of the objects in the collection and of the stories each can tell. For those who want to see more, many instruments are on display in various galleries across Royal Museums Greenwich. Details and images of these and others can be also found on the Museum's Collections Online website.[68] The Museum is always keen to encourage research on its collections, and instruments not on display can be viewed by appointment. Further work can only serve to bring to the fore the sorts of stories highlighted in this chapter and shed more light on a rich and fascinating collection of the practical tools of mathematics.

11

The University of Greenwich at the Old Royal Naval College

Noel-Ann Bradshaw and Tony Mann

CONTENTS

The authors of this chapter were both, at the time of writing in early 2018, members of the Department of Mathematical Sciences at the University of Greenwich and as such they inevitably present this part of the story of mathematics at Greenwich from the inside and do not claim objectivity.

11.1 Introduction

The former Greenwich Hospital buildings, which as we have seen became in the nineteenth century the Royal Naval College, continue to be home to mathematicians, and specifically those doing research in numerical mathematics, through the current occupiers, the University of Greenwich (Figure 11.1).

In 1995, in anticipation of the Navy's vacating the building, the then Secretary of State for Defence, Michael Portillo, had sought 'expressions of interest from organisations able to propose appropriate uses' for the buildings of the Royal Naval College. After public debate, which included an article in the *Burlington Magazine* entitled 'Greenwich Grotesquerie' which was critical of the process, new uses were found. The Greenwich Foundation for the Old Royal Naval College leased the site from the Greenwich Hospital, and two longstanding educational establishments moved into the buildings.[1] The University of Greenwich occupies King William, Queen Anne and Queen Mary Courts, along with the former Dreadnought Naval Hospital and the new Stephen Lawrence Building, and also took over the former nurses' home, now Devonport House, across the main road.

FIGURE 11.1   The Old Royal Naval College viewed from across the river. (Courtesy of the University of Greenwich.)

King Charles Court became home to Trinity College of Music (which following a merger with the Laban Dance Centre became in 2005 the Trinity Laban Conservatoire of Music & Dance).

The University of Greenwich began life in 1890 as the Woolwich Polytechnic, Britain's second polytechnic. An early charter stated, 'The object of this institution is the promotion of the industrial skill, general knowledge, health and well-being of young men and women belonging to the poorer classes ...' and in the 1930s, continuing in this tradition, the Polytechnic pioneered part-time day-release and sandwich courses[2]: the University's mathematics degrees today have a strong focus on developing students' employability skills. The Woolwich Polytechnic became Thames Polytechnic in 1970, by which time it had expanded to include various other institutions, including Avery Hill and Garnett Colleges, two teacher training establishments. The Polytechnic became the University of Greenwich in 1992. In addition to its Greenwich campus centred on the Old Royal Naval College, the University has two other campuses, at Avery Hill in south-east London and down river at Chatham.

The University's move into what is now called the Old Royal Naval College began when it occupied Queen Anne Court and the Dreadnought former hospital building in the summer of 1999. It then moved into Queen Mary and King William Courts when these became available in 2000 and early 2002, respectively. A new library and academic building in nearby Stockwell Street opened in 2015. At the time of writing the University's Faculty of Architecture, Computing and Humanities (which includes the Department of Mathematical Sciences) and Faculty of Business are based in the Old Royal Naval College. (In August 2018 the Faculty of Architecture, Computing and Humanities became the Faculty of Liberal Arts and Sciences under Pro Vice-Chancellor Professor Mark O'Thomas.)

11.2 Mathematics Research at the University of Greenwich Today

The then School of Computing and Mathematical Sciences was one of the first parts of the University to move into the new campus, first in Queen Anne Court in 1999 and then in the summer of 2000 moving into its present home in Queen Mary Court. It is a pleasing coincidence that much of the University's research work in mathematics, which has an international reputation, is very much in the numerical and computational tradition of the mathematics of the Nautical Almanac Office. This is entirely fortuitous, since the University had no previous direct connection with the establishments previously occupying its new quarters, but it shows how the methods developed at Greenwich by pioneers of computing like Maskelyne and Comrie, and the challenges they faced, had been developed and become mainstream mathematical research areas by the final quarter of the twentieth century.

The University's research work in this area was developed by Mark Cross, who had joined the then Thames Polytechnic as Professor of Computational Modelling, and subsequently became Head of the School of Mathematics, Statistics and Computing. In the 1980s this School developed a then innovative degree combining these three subjects, with a strong practical element and including a 'Modelling Week' during which the normal curriculum was abandoned as students worked in teams to tackle difficult and ill-specified real-life mathematical problems. Cross also built up research teams working on applied mathematical modelling, computational fluid dynamics and algorithms for parallel computing.

By the time mathematics at the University moved from Woolwich to Greenwich, Cross had become Pro Vice-Chancellor with responsibility for research: he subsequently moved in 2005 to Swansea University as Professor of Computational Modelling, a position in which he could focus on his own research. The Head of the School of Computing and Mathematical Sciences at the time of the move was Cross's successor, Professor Martin Everett, who worked on applications of graph theory to the social sciences. Everett moved to the University of Westminster in 2003 and was subsequently Vice-Chancellor of the University of East London: his successor as Head of School (and later Dean) at Greenwich was the computer scientist Professor Liz Bacon (who in 2014 became President of BCS – the Chartered Institute for IT, previously the British Computer Society).

At the time of the move to Greenwich the Head of Department of Mathematical Sciences was the statistician Professor Keith Rennolls. He was succeeded in 2002 by Tony Mann. Following a reorganisation in 2013 the Department of Mathematical Sciences, which had been under the leadership of Professor Kevin Parrott since 2010, came into the Faculty of Architecture, Computing and Humanities. Dr Mary McAlinden joined as Head of Department in July 2015.

Although the bulk of the mathematical research at the University continues to be in applied computational modelling, there is also work in other areas of mathematics. In pure mathematics, for example, shortly after his retirement in 2004, T.F. Tyler, who had lectured at the institution since the 1960s, completed a PhD in complex analysis at Imperial College under the supervision of Walter Hayman, who had been one of the speakers at the British Mathematical Colloquium at Greenwich in 1952 (see Chapter 6), while Neil Saunders, who joined the University in 2017, works on geometrical representation theory. Professor Vitaly Strusevich and others work in scheduling theory (his research student Kabir Rustogi won the Operational Research Society's prize for the best PhD in Operational Research in 2014); Kevin Parrott, who retired in 2015, supervised PhD work in the mathematics of finance (reflecting the University's proximity to the financial centres of the City of London and Canary Wharf (Figure 11.2)); and Keith Rennolls, who retired in 2007, led a statistics group active in fields such as medical and educational statistics.

One particular piece of work carried out in the early years of this century by Rennolls, with his doctoral student Mingliang Wang, has resonances with earlier work at Greenwich. Today's environmental scientists studying forestry and biodiversity rely on photographic images taken from satellites. These have to be matched with the true position on the ground (i.e. its precise latitude and longitude), for example when comparisons are to be made with previous images to determine changes in land-use. Rennolls and Wang worked on statistical methods to help this matching and have devised methods which give, for almost all points, an accuracy of co-registration of a fraction of a pixel.[3] The mathematicians of the past at Greenwich, searching for ways to determine the longitude of a ship at sea with the advanced technology of the seventeenth and eighteenth centuries, would surely have appreciated the mixture of mathematics and twenty-first century technology contributing to the study of the major scientific problem of today, the environmental crisis.

But it is for the breadth of its work in computational mathematics above all that the university has gained an international reputation in mathematics. This work uses computational fluid dynamics, finite element analysis and other computational techniques, with appropriate mathematical algorithms to solve the resulting equations on high-performance computers, to construct and analyse large-scale three-dimensional mathematical models of the physical world. And some of this work has a nautical theme.

FIGURE 11.2   The Queen's House and the Old Royal Naval College with Canary Wharf behind. (Courtesy of the University of Greenwich.)

Professor Ed Galea's Fire Safety Engineering Group (FSEG) uses mathematics to model how fire spreads and how this spread can be controlled, in structures such as buildings, aircraft and ships, and has extended the mathematical modelling into the behaviour of people evacuating these structures in emergencies. By helping architects design structures that can be quickly and safely evacuated this application of mathematics genuinely saves lives. FSEG's EXODUS software models people's psychology as well as their physical behaviour – which is necessary since people's reactions to fire alarms depend on all sorts of circumstances. This group's work was awarded the Queen's Anniversary Prize for Higher Education in 2002 and the British Computer Society Gold Medal for IT. The system maritime EXODUS, which is used for safety analysis in the shipping industry, has won the Communications & IT in Shipping Award for Innovation in IT for Ship Operation and Safety, and the RINA/ Lloyds Register Award for Ship Safety. FSEG was by this time a 30-strong interdisciplinary team, many of whom had been mathematics undergraduates at Greenwich: the team was shortlisted for the Times Higher Education Award for the Outstanding Engineering Research Team of the Year in 2011 and won the Guardian University Award for Research Impact in 2014.[4] Aoife Hunt, a Greenwich graduate who subsequently was employed as a Lecturer and then Senior Lecturer in the Department of Mathematical Sciences, was awarded a Student Scholar Award by the Society of Fire Protection Engineers in 2016 for her doctoral work with FSEG,[5] while a subsequent doctoral student in the group, Sandra Vaiciulyte, was selected to present a talk at the 2018 British Science Festival about her work on modelling of evacuation in the event of wildfire.

Professor Chris Bailey leads the Computational Mechanics & Reliability Group (CMRG). One high-profile project is this group's work on the conservation of the ship *Cutty Sark* (Figure 11.3). The clipper, launched at Dumbarton in 1869, was in her time one of the fastest ships in the world. She has been preserved in dry dock at

FIGURE 11.3 The *Cutty Sark*. (Courtesy of the University of Greenwich.)

Greenwich since 1954 and is now a major tourist attraction: indeed she gave her name (which comes from the young witch in Burns's *Tam O'Shanter*) to the Docklands Light Railway station for maritime Greenwich.

But by the beginning of the twenty-first century the ship was inevitably showing signs of age. She is made of wood and iron: the wood was rotting and the iron rusting, and the chemicals used to treat one were not necessarily good for the other. Furthermore the stresses caused by the support in dry dock were deforming the ship's structure. (The ship survived remarkably well a fire in May 2007: fortunately, much of the woodwork had been removed from the site during the restoration.) As part of a multi-million pound conservation project, Bailey and Stoyan Stoyanov of CMRG used mathematical modelling to find the optimal ways to treat the structure and to evaluate the different possible ways of supporting the ship. This unusual computational modelling project has helped to preserve a Greenwich icon for future visitors.[6] Bailey and his team received the Times Higher Education Award for the Outstanding Engineering Research Team of the Year in 2009.

Professor Koulis Pericleous leads the Computational Science and Engineering Group (CSEG). This group's research focuses on multi-physics modelling – simulations of systems that involve several simultaneous physical models – creating algorithms and software. It works on industrial and environmental applications involving fluid flow, heat transfer, chemical reaction, phase change, stress analysis, acoustics, and even electromagnetics, gaining an international reputation for its expertise in metals and minerals processing.

The Numerical and Applied Mathematics Unit, led by Professor Choi-Hong Lai, applies numerical mathematics to a wide range of applications, including mathematical finance, computational biology, image processing, and computational acoustics. Its work illustrates the diversity of areas in which numerical mathematics is enlarging our understanding. Examples include mathematical modelling of heart muscles in medicine, numerical solution for option pricing with non-linear volatility in the financial world and the use of mathematics in aeroacoustics to make aircraft quieter.

Other Professors in the Department of Mathematical Sciences are Valdis Bojarevics, who specialises in magnetohydrodynamics in CSEG, and Mayur Patel, who has been a member of CMRG, CSEG and FSEG, while Mike Worboys, Professor of Spatial Informatics, was for a time a member of the Department.

11.3 Public Engagement

As we have seen, Greenwich mathematicians have always considered themselves part of the wider mathematical community. Many of the people discussed in this book have held office in the Mathematical Association and its predecessor, the Association for the Improvement of Geometrical Teaching, or the London Mathematical Society, or have devoted time and energy to what we now call public engagement.

Continuing in the traditions of Greenwich mathematics discussed above, University mathematicians continue to engage with the wider mathematical community and professional bodies and to perform outreach activities. Mary McAlinden, amongst many other roles in the national mathematics education community, is currently the representative for higher education on the Advisory Committee for Mathematics Education (ACME), the influential committee set up by the Royal Society to advise on policy relating to mathematics. Tony Mann has been Newsletter Editor for the London Mathematical Society, and Noel-Ann Bradshaw is currently Vice-President (Communications) for the Institute of Mathematics and its Applications (IMA).

Bradshaw has also been active in public engagement activities for the Operational Research Society, both Bradshaw and Mann have presented events at the annual British Science Festival, and Mann has given public lectures as Visiting Professor at Gresham College. Tim Reis and Ana Paula Palacios, who joined the Department in 2016, have carried out practical demonstrations of the fluid dynamics of custard with visitors at Butlin's holiday camps. Custard is an example of a non-Newtonian fluid: it resists pressure so that it is possible to walk across a pool of custard without sinking in. Reis demonstrated this activity on the Channel 4 television programme *Food Unwrapped* in 2017.[7] Many members of the Department deliver mathematics masterclasses for school students on a wide range of topics, and Bradshaw is a presenter for the 'Citizen Maths' MOOC – a massive open online course for members of the public wishing to improve their numeracy skills. Bradshaw also initiated the Tomorrow's Mathematicians Today annual conference for undergraduate mathematicians, first held at Greenwich in 2010 and now run by the IMA[8]: this is an innovative event which gives students the opportunity to experience a national conference and present to their peers from around the country (Figure 11.4).

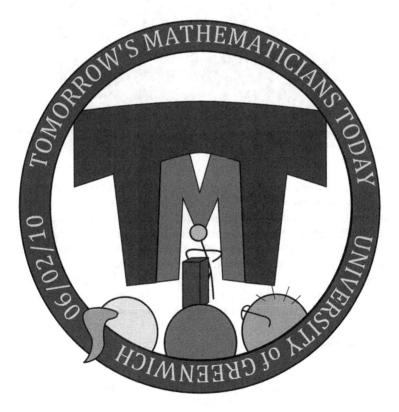

FIGURE 11.4 Logo for Tomorrow's Mathematicians Today conference, 2010. (Courtesy of Greenwich Maths Centre.)

FIGURE 11.5   Poster for the 2017 IMA Festival of Mathematics and its Applications at Greenwich. (By Lisa Fevrier, courtesy of Greenwich Maths Centre.)

In 2015 the University established the Greenwich Maths Centre, with Mann as its Director. Its role is 'to promote mathematics at all levels: to demonstrate the value of mathematics as a tool for business and industry, as an instrument for understanding our world, and as a fascinating pursuit in its own right'.[9] Its outreach activities included hosting with the IMA in June 2017 the national Festival of Mathematics and its Applications, which attracted over 1,500 visitors and presented over 70 free mathematical events for the general public (with Tim Reis offering visitors the opportunity to walk across a bath of custard) (Figure 11.5).[10] Mathematics communication has been important to Greenwich mathematicians for over four centuries.

11.4 The History of Mathematics

Given its rich mathematical history, it is perhaps not surprising that Greenwich has recent connections with the study of the history of mathematics. It was at a meeting in 1971 at Thames Polytechnic (then situated at Woolwich close to Greenwich Town Hall) that the British Society for the History of Mathematics (BSHM) was founded. The connection with the Polytechnic was through John Dubbey, a member of the mathematics department, who was to be President of the BSHM from 1977 to 1979. The department continues to have strong associations with BSHM: Mann was President from 2009 to 2011 and Bradshaw was Treasurer from 2010 to 2013, both having also held other roles on Council.

The BSHM subsequently held a meeting at the Royal Naval College in October 1997, celebrating William Burnside and marking the centenary of the publication of his *Theory of Groups of Finite Order*. The distinguished speakers included Walter Feit, whose proof with J.G. Thompson of Burnside's Conjecture about groups of odd order was mentioned in Chapter 8. (An unusual feature of this meeting was that two local artists, Vanilla Beer and Diana Pooley, were invited to record the event: examples of their work are shown in Figures 11.6 and 11.7.)

In May 2002 the BSHM had held in King William Court, in what is now the Burnside Lecture Theatre, a meeting entitled 'Understanding Mathematics through History', celebrating the life of its former President John Fauvel (1947–2001), with speakers the mathematician and writer Ian Stewart and the historian of mathematics Jan van Maanen. Other BSHM meetings at Greenwich have examined Greenwich's mathematical connections (June 2002, at the National Maritime Museum), John von Neumann (King William Court, November 2003), Maths In View (a meeting exploring the presentation of the history of mathematics on screen, which attracted about 400 visitors, in December 2008) and History of Mathematics in the Undergraduate Curriculum (King William Court, March 2010). The last of these included a presentation about a

FIGURE 11.6   Diana Pooley, M.F. Newman talking about Tarski Monsters, pastel on paper, 49 by 38 cm. (Private collection, courtesy of Diana Pooley.)

WALTER FEIT
WATCHING TONY MANN
25 IX 97

FIGURE 11.7   Vanilla Beer, Walter Feit listening to Tony Mann, pencil on paper, 16 by 18 cm. (Private collection, courtesy of Vanilla Beer.)

project involving Bradshaw and Mann from Greenwich and Mark McCartney of the University of Ulster, which created 'bite-sized' learning resources (text-based and podcasts) on the history of mathematics. This project was funded by the Higher Education Academy Mathematics, Statistics and Operational Research Network, and the products have been extensively used in subsequent teaching at Greenwich.[11]

11.5 The Teaching of Mathematics at the University

It cannot have escaped the reader that almost without exception the people mentioned in this book so far, until the twenty-first century, have been almost exclusively white and male. There are obvious historical reasons for this. But the contrast with the present-day University of Greenwich is striking. During the last few years women have in some cohorts outnumbered men amongst undergraduates reading mathematics at Greenwich and have been particularly over-represented amongst the first class honours graduates (although women are still in the minority as doctoral students and teaching staff). About 70% of students come from ethnic minorities. The University sees this as a measure of its effectiveness in widening participation to higher education amongst the local community.

As one would expect, Greenwich mathematics graduates go into a wide range of careers. Recent graduates work in the financial sector (including several employed by the Big Four accountancy companies and others working for investment banks), in data science (an increasingly attractive destination), in government statistics and pharmaceutical statistics, in logistics and operational research, for Google, Facebook and elsewhere in the IT industry, in mathematical modelling, and in teaching, to mention just some of the recent destinations. Since mathematics graduates have highly sought after skills, there are many options.

An example of a recent Greenwich graduate who has used her mathematical expertise in perhaps unexpected directions is Takita Bartlett-Lashley (Figure 11.8), who completed her degree in Financial Mathematics. Bartlett-Lashley creates mathematically inspired jewellery: her pieces have been worn at awards ceremonies by the singer and songwriter Ellie Goulding. In addition, like so many others we have come across in this book, she is passionate about promoting mathematics. She gives workshops about where maths meets fashion for school students, and (using experience gained on a final year work placement) has set up the social enterprise company STEM in Style 'providing creative learning spaces for young people to explore the Science, Tech, Engineering & Maths of Fashion and Design'.[12] Bartlett-Lashley is an example of how Greenwich mathematics graduates are using their knowledge in new

FIGURE 11.8   Takita Bartlett-Lashley with Noel-Ann Bradshaw. (Courtesy of the University of Greenwich.)

ways, while continuing the Greenwich tradition of engaging with diverse audiences in promoting their subject.

The number of mathematics undergraduates at Greenwich declined at the beginning of the twenty-first century, when the introduction of Curriculum 2000 had a disastrous effect on the uptake for Mathematics A-level in the UK, but recovered subsequently, and the Department of Mathematical Sciences now recruits about 80 new undergraduates each year. It offers programmes in mathematics, statistics and operational research, and a programme in financial mathematics was introduced in 2008: the programme leader for these is Nadarajah Ramesh. An MMath programme, under the leadership of Erwin George, was added to the portfolio in 2016. Other long-serving members of the teaching staff include Alan Soper, Yvonne Fryer, and Steve Lakin.

There is a commitment to producing graduates with strong communication and teamwork skills, and staff including Ramesh and Bradshaw have gained funding from the Higher Education Academy and the National HE STEM[13] Programme Mathematics Curriculum Innovation Fund for initiatives in this area. Providing appropriate support for students, many of whom are the first member of their family to attend university, is a priority, and Bradshaw introduced in 2012 the idea of a weekly opportunity for mathematics students and staff to meet informally to play simple strategy games and solve puzzles, while giving students the opportunity to go through with staff any material they might be finding difficult. This was immediately popular with students and, under the name 'Maths Arcade', has been taken up by around a dozen university mathematics departments across the sector (Figure 11.9).[14]

The Department scores well for student satisfaction, consistently recording high scores in the National Student Survey which asks final year undergraduates at UK universities about how satisfied they are with their courses, and it has a reputation for innovative teaching. Staff regularly present at mathematics teaching conferences, and Bradshaw and Mann have been editors of *MSOR Connections*,[15] a peer-reviewed

FIGURE 11.9   Noel-Ann Bradshaw with a student at the Maths Arcade. (Courtesy of the University of Greenwich.)

online mathematics education journal hosted by the University since 2015, and members of the steering group of the **sigma** network, which encourages the provision of support in mathematics and statistics for university students whatever their subject.[16]

The University, in all its various incarnations, has always been committed to education for practitioners, with a strong widening participation ethos, and the University's degree programmes have a strong focus on the employability of its graduates.

One particularly popular option for final-year mathematics students at Greenwich is to take part in the Undergraduate Ambassadors Scheme.[17] This national scheme was initially devised and funded by the author and broadcaster Simon Singh: Greenwich was one of the universities which enthusiastically adopted it early on. A significant proportion of the university's final-year maths students now spend a day each week helping mathematics teachers in a local school or college. Many of these undergraduates are inspired to enter the teaching profession after graduation, helping address the national shortage of mathematics teachers. In their work in schools, which they report to be a thoroughly enjoyable and rewarding experience, they are serving as role models and helping to educate the next generation of Greenwich mathematicians.

11.6 Conclusion

This book has told the story of mathematics, and especially numerical mathematics and computation, at Greenwich from 1675 (Figure 11.10). It has shown mathematics in the service of navigation, from the longitude problem to the work of the

FIGURE 11.10   The Old Royal Naval College in snow, with the Queen's House and the Observatory behind. (Courtesy of the University of Greenwich.)

Nautical Almanac Office in the twentieth century. We have seen mathematics helping us develop our understanding of the cosmos in which we live, and helping with the development of first Newton's and then Einstein's models. We have covered the significant work on numerical computation of mathematicians like John Flamsteed, Nevil Maskelyne and Leslie Comrie. It is satisfying that mathematics in these traditions continues at Greenwich today, finding new applications in the modern world but very much in the spirit of the pioneers discussed in this book.

12

The Mathematical Tourist in Greenwich

Tony Mann

Greenwich has much to offer any visitor. The Royal Park, the National Maritime Museum, the Cutty Sark, Greenwich Market, and many other attractions are popular destinations for tourists, who can travel from central London conveniently by train or Docklands Light Railway, or more picturesquely by river. But Greenwich has a number of specifically mathematical points of interest.

The Observatory has featured throughout this book. Situated at the top of Greenwich Park, it is now part of Royal Museums Greenwich, along with the National Maritime Museum and the Queen's House. Visitors can tour Flamsteed House, including the historic Octagon Room (illustrated in the Introduction to this volume), and there are displays illustrating the history of time and including Harrison's historic chronometers.

At the foot of the hill the National Maritime Museum contains the objects described by Richard Dunn in Chapter 10 and many other fascinating exhibits. The Queen's House was designed originally by Inigo Jones for Queen Anne of Denmark, wife of King James VI and I, before her death in 1619, and completed for Queen Henrietta Maria in 1635, with proportions based on mathematical principles. Now it displays art from the Maritime Museum's collections and temporary exhibitions.

Details of opening hours and admission charges for Royal Museums Greenwich can be found at www.rmg.co.uk/

Continuing towards the river, we pass Devonport Mausoleum which currently contains the monument to Edward and John Riddle described by Bernard de Neumann in Chapter 5. The Mausoleum is open to the public only during London's Open House weekend (www.londonopenhouse.org/) (during which other Greenwich buildings which are not normally open to the public are also accessible).

The buildings of the Old Royal Naval College now house the University of Greenwich and Trinity Laban Conservatoire of Music and Dance. The Painted Hall and the ornate Chapel are open to visitors, the Painted Hall having reopened in 2019 after major conservation work. James Thornhill's spectacular paintings on the walls and ceiling of the Painted Hall, on which the artist worked between 1708 and 1726, contain a number of mathematical references, especially on the wall at the lower end of the Hall (looking back towards the entrance).

> On the Left Hand is that noble Danish Knight *Tycho Brahe*, near him is *Copernicus* with his SYSTEM in his Hand, by him is an old Philosopher pointing to some remarkable Mathematical Figures of the incomparable Sir *Isaac Newton*.

On the Right in this Gallery is the celebrated *English* Astronomer the Reverend Mr *Flamsteed*, who holds the Construction of the great Eclipse which happen'd *April* the 22d, 1715. Close by him is his ingenious Disciple *Mr Thomas Weston*, who is now Master of the *Academy* in *Greenwich*: He is assisting *Mr Flamsteed* in making Observations, with a large Quadrant (whilst an old Man at the Clock is counting the Time) of the Moon's Descent upon the SEVERN, which at certain Times when he is in her *Perigee*, makes such a Roll of the Tides, call'd the *Eagre*, as is very dangerous to all in its Way.[1]

For the opening hours of the Painted Hall and the Chapel, see www.ornc.org/

The buildings of the Old Royal Naval College, designed by the leading architects of the early seventeenth century to the site design of Sir Christopher Wren, and including James 'Athenian' Stuart's Dreadnought Hospital (now the focus for the University's student services) are also open to the public during Open House Weekend. It has been suggested that Wren's spatial design for the buildings, symmetrically framing the Queen's House when viewed from the river, embodies Newtonian ideas:

> Wren's design represents the dynamic 'will-to-infinity' through an application in mortar and stone of mathematical principles bodied forth in *Principia*, the traditional iconography of the heavenly dome taking on, in the form of Wren's Observatory, the added symbolism of the celestial science of astronomy which had been singularly advanced by Newton.[2]

The University's School of Computing and Mathematical Sciences maintains a small 'Computer Museum' in King William Court.[3] This contains a variety of old computer equipment and calculating machines. It testifies to the magpie instincts of some of the academics and technicians at Greenwich, who climbed into skips and scavenged through cellars, to preserve relics which cluttered offices and garages to the dismay of colleagues and partners. The Computer Museum has also benefited from the generosity of a number of visitors who have been inspired to offer additional exhibits. The Computer Museum is open to visitors by appointment – please email maths@gre.ac.uk for more information.

The West Gate of the Old Royal Naval College was originally in 1751 situated much closer to King Charles and King William Courts and was moved to its current position in 1850.[4] The gate is surmounted by two six-foot-diameter globes, one celestial and one terrestrial. The southern, terrestrial sphere originally had lines of latitude inlaid in copper and marked the route of Anson's circumnavigation of the globe in 1740–1744, while the northern sphere had inlays of '24 meridians, the equinoxial, ecliptic, tropics, and circles'[5] with many stars marked. They were planned by Richard Oliver, a former mathematics master at the Academy of Greenwich. The globes are now much eroded.

As one leaves the Old Royal Naval College and continues towards the river, one comes to the *Cutty Sark*, now resplendent in her new display. We saw in Chapter 11 how mathematics has contributed to the survival and preservation for the future of this icon of maritime history. She provides a fitting end to this account of the mathematical history of Greenwich.

Endnotes

Introduction

1 Tate, 'Canaletto, *A View of Greenwich from the River*' [www.tate.org.uk/art/artists/canaletto-2302, accessed 24 December 2018].

2 Royal Museums Greenwich, 'Canaletto, *Greenwich Hospital from the North Bank of the Thames*' [http://collections.rmg.co.uk/collections/objects/13306.html, accessed 24 December 2018].

3 C. Andrew Murray, 'Pond, John (*bap.* 1767, *d.* 1836)', Oxford Dictionary of National Biography, Oxford University Press, 2004 (revised 2005) [www.oxforddnb.com/view/article/22490, accessed 24 December 2018].

Chapter 1

1 See *The 'Preface' to John Flamsteed's 'Historia Coelestis Britannica'*, edited and introduced by Allan Chapman, based on a translation by Alison Dione Johnson (National Maritime Museum Greenwich Monographs and Reports No. 52 (1982)), 222 pages.

2 Chapman, *'Preface'* (n. 1), 55–60.

3 Chapman, *'Preface'* (n. 1), 61–111, for Flamsteed's full treatment of Tycho's work. For Tycho's own account of his instruments and methods, see Tycho Brahe, *Astronomiae Instauratae Mechanica* (Wandesburgi, 1598), translation by Hans Raeder, Elis Strömgren, and Bengt Strömgren as *Tycho Brahe's Description of his Instruments and Scientific Work* (Copenhagen, 1946). Also John Louis Emil Dreyer, *Tycho Brahe. A Picture of Scientific Life and Work in the Sixteenth Century* (1890), reprinted N.Y. 1963. For Hevelius's instruments, see Johannes Hevelius, *Machina Coelestis I* (Danzig, 1673). For a modern study of Hevelius, see Allan Chapman, 'Johannes Hevelius (1611–1687): Instrument Maker, Lunar Cartographer, and Surveyor of the Heavens', in Sir Patrick Moore and Dr John Mason, *Yearbook of Astronomy 2013* (Macmillan, London, 2012). Also Allan Chapman, *Dividing the Circle. The Development of Critical Angular Measurement in Astronomy, 1500–1850* (Praxis Publishing, John Wiley, Chichester, 1990, 1995), 16–33.

4 Flamsteed's career is placed in context in Jim Bennett's chapter 'Flamsteed's Career in Astronomy: Nobility, Morality, and Public Utility', in the excellent volume of essays *Flamsteed's Stars. New Perspectives on the Life and Work of the First Astronomer Royal*, ed. Frances Willmoth (Boydell Press, National Maritime Museum, London, 1997), 17–30. Also Chapman, *'Preface'* (n. 1), 105–9; *Dividing the Circle* (n. 3), 38–40.

5 Chapman, *'Preface'* (n. 1). Flamsteed makes many references to Gascoigne, Horrocks, and Crabtree and the Townleys in his early correspondence: see Letters 62–5 (1671), reproduced in *The Correspondence of John Flamsteed, First Astronomer Royal 1, 1666–1682*, eds. Eric G. Forbes, Lesley Murdin, and Frances Willmoth (Institute of Physics, 1995), 100–5, etc. Chapman, *Dividing the Circle* (n. 3), 35–45, 174–6.

6 For Gascoigne's sextant 'carkasse', see Flamsteed to Samuel Molyneux, 10 May 1960, Letter 622 in *The Correspondence of John Flamsteed, First Astronomer Royal 2, 1682–1703*, eds. Eric G. Forbes, Lesley Murdin, and Frances Willmoth (I.O.P., Bristol and Philadelphia, 1997), 418ff. Also Francis Baily, *An Account of the Revd. John Flamsteed* (C.U.P., London, 1835). Chapman, *Dividing the Circle* (n. 3), 36–7.

7 Richard Townley, 'An extract of a letter by Mr Richard Towneley [*sic*] to Dr Croon', *Philosophical Transactions of the Royal Society* 2 (1667), 458.

8 Johannes Hevelius, *Mercurius in sole visus* (Dantzig, 1662), which included the first publication of Horrocks's Venus observations, *Venus in sole visa*, 24 Nov. 1639, from a manuscript copy. Peter Aughton, *The Transits of Venus. The Brief, Brilliant Life of Jeremiah Horrocks, Father of British Astronomy* (Weidenfeld and Nicolson/ Windrush Press, London, 2004), 4–6. Chapman, 'Johannes Hevelius (1611–1687)' (n. 3).

9 Eric G. Forbes, *Greenwich Observatory 1: Origins and Early History (1675–1835)* (Taylor and Francis, London, 1975), 18–24.

10 Chapman, *'Preface'* (n. 1), 111–13. Forbes, *Greenwich Observatory 1* (n. 9), 25–46.

11 Christiaan Huygens, *Horologium* (Hague, 1658). For an English translation, by Ernest L. Edwards, see *Antiquarian Horology* 7, 1 (Dec. 1970), 38–55. Derek Howse, 'The Tompion clocks at Greenwich and the dead beat escapement, Pt. I', *ibid.*, 24–9. Derek Howse, *Greenwich Observatory 3: The Buildings and Instruments* (Taylor and Francis, London, 1975), 25–6.

12 Howse, *Greenwich Observatory 3* (n. 11), 1–6.

13 *Correspondence of John Flamsteed 2* (n. 6), Letters 597, 599 (1688), pp. 381–3, 385–6.

14 Baily, *An Account of . . . Flamsteed* (n. 6), 316. W. H. Quarrell and Margaret Ware, *London in 1710* (1934), 22: a translation of Baron Zacharias Conrad von Uffenbach's *Merkwürdige Reisen durch . . . und Engelland* (1753).

15 Chapman, *'Preface'* (n. 1), 113–20. Howse, *Greenwich Observatory 3* (n. 11), 17–19. Chapman, *Dividing the Circle* (n. 3), 54–6, plates 18, 19.

16 Letter, Flamsteed to Sir Jonas Moore, 16 July 1678: Letter 341 in *Correspondence of John Flamsteed 1* (n. 5), 643–6.

17 Howse, *Greenwich Observatory 3* (n. 11), 125–6. Howse, 'The Tompion clocks at Greenwich' (n. 11), 24–9.

18 Chapman, *'Preface'* (n. 1), 120–4ff. Howse, *Greenwich Observatory 3* (n. 11), 119–21. Chapman, *Dividing the Circle* (n. 3), 54–9, plates 20–3.

19 Chapman, *'Preface'* (n. 1), 151. Howse, *Greenwich Observatory 3* (n. 11), 126.

20 Chapman, *'Preface'* (n. 1), 175–7, 189ff. For a modern translation of the *'Preface'* to the 1712 'pirated' edition by Mrs E. Barber see above, 189–94.

21 Chapman, *'Preface'* (n. 1), 155; see also p. 21, n. 49.

22 Chapman, *'Preface'* (n. 1), 160–80, for details of how Sir Isaac Newton attempted to remove the publication of Flamsteed's work into his own control, and how Flamsteed resisted him.

23 Derek Howse, *Nevil Maskelyne: the Seaman's Astronomer* (C.U.P., New York, 1989), 201–2. Letter from Sir Joseph Banks, P.R.S., 8 January 1810.

24 George Biddell Airy, *Autobiography of Sir George Biddell Airy*, ed. Wilfred Airy (C.U.P., Cambridge, 1896). See also letters from Airy to Lord Auckland, 8 October 1832, Royal Greenwich Observatory 6 (hereafter RGO 6), 1/151–2; Airy to Duke of Sussex, 19 May 1834, RGO 6 1/145; Airy to Lord Auckland, 10 Sept. 1834, RGO 6 1/153, etc. (The RGO 6 archive, along with the papers of the other Astronomers Royal, is now incorporated into Cambridge University Library.)

25 The true extent of Halley's personal involvement with meeting the costs of printing *Principia* is unclear, but in 1686 the Royal Society was nearly bankrupt and, as Clerk to the Society, Halley was essentially told to see to the publication from his own resources: see Sir Allan Cook, *Edmond Halley. Charting the Heavens and the Seas* (Clarendon Press, Oxford, 1998), 154–5.

26 Howse, *Greenwich Observatory 3* (n. 11), 32–4, plate 17. Chapman, *Dividing the Circle* (n. 3), 83–7, plate 26, for description of how transits were used.

27 Howse, *Greenwich Observatory 3* (n. 11), 21–4. Chapman, *Dividing the Circle* (n. 3), 66–71, plate 29.

28 Edmond Halley, *A Description of the Passage of the Shadow of the Moon over England . . . 22 April 1715 . . .* Broadsheet describing the path of the forthcoming eclipse issued by Halley. Copies currently in the Royal Society and the Royal Astronomical Society Libraries. The observations of people who sent their results to Halley were published in *Phil. Trans.* See Chapman, 'The people's eclipse of 1715', *Astronomy Now*, July 1999, 23–5.

29 John Harrison listed the names of some of those who had given him financial assistance over the years – generally undated – in his *A Description Concerning such Mechanism as will Afford a Nice, or True Mensuration of Time* (London, 1775), 21. See also William J. H. Andrewes, 'Even Newton could be wrong: the story of Harrison's First Three Sea Clocks', in *The Quest for the Longitude*, ed. Andrewes (Collections of Historical Scientific Instruments, Harvard, MA, 1996), 199, notes 38, 39 ff.; p. 208 for Board of Longitude £500 payment of 1737. The leading biography on Harrison is still Humphrey Quill, *John Harrison, the Man who Found Longitude* (London, 1966).

30 Stephen P. Rigaud, 'Some particulars respecting the principal instruments at . . . Greenwich in the time of Dr Halley', *Monthly Notices of the Royal Astronomical Society* 9 (1836), 205–7. See also James Bradley, *Miscellaneous Works and Correspondence of the Rev. James Bradley*, ed. Stephen P. Rigaud (O.U.P., Oxford, 1832), 381–3, etc., for his analyses of Halley's instrument errors. Chapman, *Dividing the Circle* (n. 3), 92–7. By far the most thorough-going scholarly study of James Bradley's life and work to date is the doctoral thesis of Dr John Roger Fisher, 'The Work of James Bradley, Third Astronomer Royal of England' (Imperial College, University of London, 2004).

31 Robert Hooke, *An Attempt to Prove the Motion of the Earth* (London, 1674). Flamsteed had also tried a zenith sector with his 1676 'Well Telescope': Howse, *Greenwich Observatory 3* (n. 11), 58–60. Chapman, *Dividing the Circle* (n. 3), 87–9.

32 James Bradley, 'A letter to Dr Halley giving account of a new discovered motion of the fixed stars', *Phil. Trans.* 35 (1728), 637–61.

33 James Bradley, 'A letter to the Earl of Macclesfield giving Account of a new discovered motion of the fixed stars', *Phil. Trans.* 45 (1748), 1–36. Bradley's work on the aberration and nutation are also discussed in his *Miscellaneous Works* (n. 30).

34 *The Herschel Chronicle. The Life-Story of William Herschel and His Sister Caroline Herschel*, ed. Constance A. Lubbock (C.U.P., Cambridge, 1933), 1–58, reprints many of William's early letters and writings that chart his life. See also

Michael Hoskin's splendid modern scholarly studies of the Herschels, such as *The Herschel Partnership, as Viewed by Caroline* (Science History Publications, Cambridge, 2003), and *Caroline Herschel's Autobiographies* (Science History Publications, Cambridge, 2003).

35 David J. Bryden, *James Short and His Telescopes* (Royal Scottish Museum, Edinburgh, 1968), 7, for Short's estate valuation. See also Tristram Clarke, 'James Short', *Oxford Dictionary of National Biography* (Oxford, 2004).

36 Bradley, *Miscellaneous Works* (n. 30), lv. John Bird, *A Method of Constructing Mural Quadrants* (W. Richardson and S. Clark, London, 1768), 7. The error was variously ascribed to pivot wear and to thermal distortion between the brass and iron parts.

37 Abraham Rees, *The Cyclopaedia, or Universal Dictionary of Arts, Sciences, and Literature* (1819–1820), article 'Pendulum', 237–8: reproduced in facsimile in *Rees' Clocks, Watches, and Chronometers* (David and Charles Reprints, Newton Abbot, 1970), same pagination.

38 Bradley, *Miscellaneous Works* (n. 30), lxxiv. Robert T. Gunther, *Early Science in Oxford, II* (Oxford, 1923), 320.

39 Howse, *Greenwich Observatory 3* (n. 11), 24.

40 Bradley, *Miscellaneous Works* (n. 30), viii. Chapman, *Dividing the Circle* (n. 3), 94–5. Dr Roger Fisher provides an excellent account of Bradley's instrumental procedures at Greenwich in Ch. 5 of his doctoral thesis, including Bradley's concern with quantifying and compensating for errors, and his use of inter-instrument cross-checking, with many citations from Bradley's observing logs and notes. See especially 201–15.

41 Friedrich Wilhelm Bessel, *Fundamenta Astronomia per pro anno mdcclv deducta observationibus James Bradley, . . . 1750–1760 . . .* (Regiomontani, 1818), 26ff. Allan Chapman, 'Pure research and practical teaching: the astronomical career of James Bradley 1693–1762', *Notes Rec. R. Soc.* 47 (2) (1993), 205–12.

42 Allan Chapman, 'The accuracy of angular measuring instruments used in astronomy between 1500 and 1850', *J. Hist. Astron.* 14 (2) (1983), 133–7: see graph, 134. Reprinted in Chapman, *Astronomical Instruments and their Users: Tycho Brahe to William Lassell* (Variorum Collected Studies, Aldershot, 1996), No. II.

43 Howse, *Nevil Maskelyne* (n. 23), 14. Eric G. Forbes, *The Euler-Mayer Correspondence (1751–1755)* (American Elsevier, New York, 1971), 17–20. Bruce Chandler, 'Longitude in the Context of Mathematics', in *The Quest for Longitude*, ed. Andrewes (n. 29), 38–9.

44 The original account of Hadley's 'reflecting quadrant', or octant, is given by him in John Hadley, 'Description of a new instrument for taking Angles', *Phil. Trans.* 37 (1731), 147–57. Also Alan Stimson, 'The longitude problem: The Navigator's Story', in Andrewes (ed.), *The Quest for Longitude* (n. 29), 72–84.

45 See Jesse Ramsden, *Description of an Engine for Dividing Mathematical Instruments* (London, 1777). Edward Troughton, 'An account of a method of dividing astronomical instruments by ocular inspection', *Phil. Trans.* 99 (1809), 112. David Brewster, article 'Graduation' in *Edinburgh Encyclopaedia*, X (Edinburgh, 1830). Chapman, *Dividing the Circle* (n. 3), 123–7.

46 Howse, *Nevil Maskelyne* (n. 23), 85–99. Also *Man is Not Lost. A Record of Two Hundred Years of Astronomical Navigation with the Nautical Almanac, 1767–1967* (National Maritime Museum and Royal Greenwich Observatory, H.M.S.O., 1968). An excellent 42-page booklet describing the practical usage of the Lunars Method.

For discussion of the early octant and sextant trade, see Jim Bennett, *The Divided Circle. A History of Instruments for Astronomy, Navigation, and Surveying* (Phaidon, Christie's, 1987), 130–42.

47 Anita McConnell, 'Nathaniel Bliss', in *Oxford Dictionary of National Biography* (Oxford, 2004).

48 *Journals of Captain James Cook* I, ed. John C. Beaglehole (Hakluyt Society, 1955), 448. Also in other editions of Cook's voyages for late January 1771. See also John C. Beaglehole, *The Life of Captain James Cook* (Hakluyt Society, 1974), 262, 264. Derek Howse, 'Charles Green', *Oxford Dictionary of National Biography* (Oxford, 2004).

49 For Bliss, see also Gunther, *Early Science in Oxford. II* (n. 38), 'Astronomy', 86–7.

50 Allan Chapman, 'Thomas Hornsby and the Radcliffe Observatory', in *Oxford Figures. 800 Years of the Mathematical Sciences*, eds. John Fauvel, Raymond Flood, and Robin Wilson (O.U.P., Oxford, 2000; 2nd edn, 2013), 168–85: 172 for instrument costs, 175 for Thomas Bugge's 1777 visit, with references (Pagination refers to the first edition).

Chapter 2

1 Royal Warrant 8 February 1765. Quoted in D. Howse, N. Maskelyne, (1732–1811), *Oxford Dictionary of National Biography*, Oxford: Oxford University Press, 2004–2006.

2 A. Van Helden, 'Longitude and the Satellites of Jupiter', pp. 85–100 in W.J.H. Andrewes, *The Quest for Longitude*, Cambridge, MA: Harvard University Press, 1996.

3 S. Harris, *Sir Cloudesley Shovell: Stuart Admiral*, Staplehurst: Spellmount, 2001.

4 D. Howse, *Greenwich Time and Longitude*, London: Philip Wilson, 1997, pp. 57–58.

5 For further details about the relationship between time and longitude see D. Howse, *Greenwich Time and Longitude*, London: Philip Wilson, 1997.

6 J. Werner, *Nova translatio primi libri geographic C. Ptolomaeic*, Nürnberg: Stuchs, 1514.

7 For example see
   R. T. Gould, *The Marine Chronometer: Its History and Development*, London: Holland, 1923.
   H. Quill, *John Harrison, the Man Who Found Longitude*, London: Baker, 1966.
   J. Betts, *Harrison*, London: National Maritime Museum, 1993.
   W. Andrews, ed., *The Quest for Longitude*, Cambridge, MA: Harvard University Press, 1996.
   D. Sobel, *Longitude*, London: Forth Estate, 1996.
   'Longitude' a television programme made by Granada Television and shown on UK Channel 4 on 2 and 3 January 2000.

8 A member of the Royal Society, *An Account of the Proceedings, in order to the Discovery of Longitude at Sea ...*, London: T. and J. W. Pasham, 1763; J. Harrison, *A Narrative of the Proceedings relative to the Discovery of the Longitude at Sea*, London: Sandby, 1765; J. Harrison, *Remarks on a Pamphlet Lately Published by the Rev. Mr. Maskelyne, under the Authority of the Board of Longitude*, London, 1767.

9 D. Howse, *Nevil Maskelyne: The Seaman's Astronomer*, Cambridge: Cambridge University Press, 1989.

10 E.G. Forbes, 'Tobias Mayer (1723–1762): A Forgotten Genius', *British Journal for the History of Science*, vol. 5, no. 17, 1970, pp. 1–20.
E.G. Forbes, 'Tobias Mayer's Lunar Tables', *Annals of Science*, vol. 22, no. 2, 1966, pp. 105–116.

11 N. Maskelyne, *The British Mariner's Guide Containing Complete and Easy Instructions for the Discovery of Longitude at Sea*, London, 1763.

12 N. Maskelyne, *The British Mariner's Guide Containing Complete and Easy Instructions for the Discovery of Longitude at Sea*, London, 1763, pp. iv–v.

13 D. Howse, *Nevil Maskelyne: The Seaman's Astronomer*, Cambridge: Cambridge University Press, 1989, pp. 40–52.

14 Minutes of the Board of Longitude. 9 Feb 1765 University of Cambridge, Manuscripts, RGO 14/5.

15 Minutes of the Board of Longitude. Sept. 1766, University of Cambridge, Manuscripts, RGO 14/5 folio 113.

16 Minutes of the Board of Longitude. Dec. 1767, University of Cambridge, Manuscripts, RGO 14/5.

17 A full discussion of Maskelyne's work computing the Nautical Almanac is given in M. Croarken, 'Tabulating the Heavens: Computing the Nautical Almanac in 18th-century England', *IEEE Annals of the History of Computing*, vol. 25, no. 3, 2003, pp. 48–61.

18 N. Maskelyne, Diary of Nautical Almanac Work, Cambridge University Library, Manuscripts, RGO 4/324, folio 42.

19 Cambridge University Library, Manuscripts, RGO 4/216.

20 Papers of Joshua Moore, MMC-2121, Library of Congress Manuscript Division, Washington, DC.

21 M. Hitchins to J. Moore, 9 August 1788, Papers of Joshua Moore, MMC-2121, Library of Congress Manuscript Division, Washington, DC.

22 Diary of Nautical Almanac Work, Cambridge University Library, Manuscripts, RGO 4/324, folio 71–79.

23 Minutes of the Board of Longitude 13 January 1770. University of Cambridge, Manuscripts, RGO 14/5, folio 188.

24 Minutes of the Board of Longitude. University of Cambridge, Manuscripts, RGO 14/5.

25 Diary of Nautical Almanac Work, Cambridge University Library, Manuscripts, RGO 4/324.

26 Papers of Joshua Moore, MMC-2121, Library of Congress Manuscript Division, Washington, DC.

27 For example Hutton and Wales have Dictionary of National Biography entries which barely mention their Nautical Almanac employment.

28 See M. Croarken, 'Mary Edwards: Computing for a Living in 18th-century England', *IEEE Annals of the History of Computing*, vol. 25, no. 4, 2003, pp. 9–15.

29 D. Howse, 'Maskelyne, Nevil (1732–1811)', *Oxford Dictionary of National Biography*, Oxford University Press, 2004–2006.

30 An account of the life and work of Maskelyne's assistants is given in M. Croarken, 'Astronomical Labourers: Maskelyne's Assistants at the Royal Observatory, Greenwich, 1765–1811', *Notes and Records of the Royal Society*, 2003, vol. 57, no. 3, pp. 285–298.

31 N. Maskelyne, *Astronomical Observations Made at the Royal Observatory at Greenwich*, London: J Nourse and others, 1776, 1787, 1799, 1811.

32 S.L. Chapin, 'Lalande and the Longitude: A Little Known London Voyage of 1763', *Notes and Records of the Royal Society*, vol. 32, no. 2, 1978, pp. 165–180.

33 D. Howse, *Nevil Maskelyne: The Seaman's Astronomer*, Cambridge: Cambridge University Press, 1989, p. 90.

34 D. Grier, *When Computers Were Human*, Princeton, NJ: Princeton University Press, 2005, p. 55.

35 S.J. Dick, 'History of the American Nautical Almanac', pp. 11–53 in A.D. Fiala and S.J. Dick, *Proceedings Nautical Almanac Office Sesquicentennial Symposium, U.S. Naval Observatory March 3–4, 1999*, Washington, DC: U.S. Naval Observatory, 1999; C.H. Davis, *Remarks on an American Prime Meridian*, Washington, DC, 1849.

36 D. Howse, *Greenwich Time and the Longitude*, London: Philip Wilson, 1997, p. 128.

37 U.S. Government, *International Conference held in Washington for the Purpose of Fixing a Prime Meridian and a Universal Day, October 1884 – Protocols of the Proceedings*, Washington, DC, 1884.

38 Diary of Nautical Almanac Work, Cambridge University Library, Manuscripts, RGO 4/324.

Chapter 3

1   Chapman, Allan. *Gods in the Sky: Astronomy, Religion and Culture from the Ancients to the Renaissance*. London: Channel 4 Books, 2001.

2   Adams, John Couch. 'History of the Discovery of Neptune.' *Memoirs of the Royal Astronomical Society* 16 (1847): 385–414.

3   The exact date of Adams' visit is unknown but thought to be 21 September 1845. See Chapman, Allan. 'Private Research and Public Duty: George Biddell Airy and the Search for Neptune.' *Journal for the History of Astronomy* 19.57 (1988): 121–39.

4   Bollobás, Béla, ed. *Littlewood's Miscellany*. Cambridge: Cambridge University Press, 1986, p. 174, *et seq.* for Littlewood's reworking of the analysis.

5   Airy, Wilfrid, ed. *Autobiography of Sir George Biddell Airy*. Cambridge: Cambridge University Press, 1896, p. 2. Henceforth *Autobiography*.

6   Ashworth, William J. 'The Calculating Eye: Baily, Herschel, Babbage and the Business of Astronomy.' *British Journal for the History of Science* 27 (1994): 409–41, p. 424.

7   *Autobiography*, p. 152.

8   For Airy's professional standing see Chapman, Allan. 'Britain's First Professional Astronomer.' *Yearbook of Astronomy*. Ed. Patrick Moore. London: Sidgwick and Jackson, 1992, pp. 185–205. Chapman, Allan. 'Science and the Public Good: George Biddell Airy (1801–1892) and the Concept of a Scientific Civil Servant.' *Science, Politics and the Public Good: Essays in Honour of Margaret Gowing*. Ed. Nicolaas A. Rupke. London: Macmillan, 1988, pp. 36–62.

9   See for example Hyman, Anthony. *Charles Babbage: Pioneer of the Computer*. Oxford: Oxford University Press, 1982, p. 191.

10 Babbage, Charles. *The Exposition of 1851 or Views on the Industry, the Science, and the Government of England*. 2nd ed. London: John Murray, 1851. Reprinted in: Campbell-Kelly, Martin, ed. *The Works of Charles Babbage*. Vol. 10. London: William Pickering, 1989. Henceforth *Works*.

11 The quoted phrase is by Dionysus Lardner. See note 12, *Works*, Vol. 2, p. 169.

12 Lardner, Dionysus. 'Babbage's Calculating Engine.' *Edinburgh Review* 59 (1834): 263–327. Reprinted in Works, Vol. 2, pp. 118–86. For the detailed circumstances surrounding the publication of Lardner's article see Swade, Doron. 'Calculation

and Tabulation in the 19th Century: George Biddell Airy Versus Charles Babbage.' PhD, University College London, 2003, Chapter 3. Babbage's Expectations for his Engines.

13 For a detailed analysis of the five papers and of the utility of the engines, see Swade, note 12, Chapter 3.

14 Swade, Doron. 'Calculating Engines: Machines, Mathematics, and Misconceptions.' *Mathematics in Victorian Britain.* Eds. Flood, Raymond, Adrian Rice and Robin Wilson. Oxford: Oxford University Press, 2011, pp. 238–59.

15 Swade, Doron. 'The "Unerring Certainty of Mechanical Agency": Machines and Table Making in the Nineteenth Century.' *The History of Mathematical Tables: From Sumer to Spreadsheets.* Eds. Campbell-Kelly, Martin, Mary Croarken, Raymond Flood and Eleanor Robson Oxford: Oxford University Press, 2003, pp. 143–74.

16 Hyman, Anthony. *Charles Babbage: Pioneer of the Computer.* Oxford: Oxford University Press, 1982, pp. 193–5.

17 Peel to Buckland, 31 August 1842. BL. Add Ms 40514, ff. 223–4. Quoted in Swade, Doron. *The Cogwheel Brain: Charles Babbage and the Quest to Build the First Computer.* London: Little, Brown, 2000, p. 135. Italics original.

18 For discussion see note 12, 'Calculation and Tabulation', p. 197.

19 The Royal Society Engine Committees reported in 1823, 1829, and 1830. For membership see Weld, Charles Richard. *A History of the Royal Society.* 2 vols. London, 1849. Reprinted in *Works*, Vol. 10, pp. 151, 153, 157.

20 See Swade (2000), pp. 52–4.

21 Goulburn to Airy, 15 September 1842, RGO6-427, f. 63.

22 Airy to Goulburn, 16 September 1842. RGO6-427, f. 65.

23 Thomas Young (1773–1829) was Superintendent of the Nautical Almanac from 1818 to 1829. He was also Secretary of the Board of Longitude. For a biography of Young see Robinson, Andrew. *The Last Man Who Knew Everything.* Oxford: One World, 2006.

24 Young's objection is recorded in Weld (1849), *Works*, Vol. 10, p. 151, note 5.

25 Airy to Goulburn, 16 September 1842. RGO6-427, f. 65. Emphasis original.

26 *Autobiography*, p. 4.

27 *Ibid.*

28 *Ibid.*, pp. 12, 1.

29 *Ibid.*, p. 12.

30 Herschel to Airy, 27 September 1842. RGO6-427, f. 74.

31 Airy to Goulburn, 16 September, 1842. RGO6-427, f. 65.

32 For quoted phrase see Peel to Buckland, 31 August 1842. BL Add Ms 40514, ff. 223–4.

33 Goulburn to Airy, 17 September 1842. RGO6-427, f. 67.

34 For membership of the Committee see Weld (1849). *Works*, Vol. 10, p. 151. For dates of Airy's undergraduate time at Cambridge see W. Airy (1896), p. 12.

35 Babbage's Diary 1820–1825, Waseda University Library. The first entry on the RS Committee is dated 21 April [1823], the date of the first meeting. The three days before the first meeting (18, 19 and 20 April) were set aside for members to examine the engine at Babbage's house. It is during these inspection visits that Babbage would have had an opportunity to make his case.

36 For a fuller account of Young's position see note 12, 'Calculation and Tabulation', pp. 205–8. Also see below for the substance of Young's objection.

37 For Airy's cryptic reference to legal proceedings during the Lucasian election see *Autobiography*, p. 70. For an account of the 1826 contest see, Schaffer, Simon. 'Paper and Brass: The Lucasian Professorship 1820–1839.' *From Newton to Hawking: A History of Cambridge University's Lucasian Professors of Mathematics*. Eds. Knox, Kevin C. and Richard Noakes. Cambridge: Cambridge University Press, 2003.

38 Goulburn to Babbage, 3 November 1842. BL Add Ms 37192 f. 172. The letter is reprinted in full in *Works*, Vol. 10, pp. 161–2.

39 *Autobiography*, p. 152.

40 *Ibid.*, p. 66.

41 *Ibid.*, p. vii.

42 *Ibid.*, p. 69.

43 T. R. Robinson to Babbage, 20 December [1835], Armagh Observatory. BL Add Ms 37189, f. 220. The year is inferred from the sequence before and after the relevant folio.

44 Toynbee, William, ed. *The Diaries of William Charles Macready 1833–1851*. 2 vols. London: Chapman and Hall, 1912. 17 September 1837. Vol. I, p. 410.

45 Airy to Goulburn, 16 September 1842, RGO6-427, f. 65.

46 Weld (1849). *Works*, Vol. 10, p. 151, ft. 5.

47 As a rough guide the estimated total costs of the Difference Engine if completed were £35,000. Interest rates were stable at 3%. Computers were paid about £120 per annum. As an indicative guide of costs Airy recorded that the cost of correcting 2,560 astronomical observations made between 1830 and 1853 which were in error following a mistake found in Burckhardt's formula for parallax correction was £400. See *Autobiography*, p. 214.

48 *Autobiography*, p. 36.

49 *Ibid.*, p. 40.

50 Peacock, George. *Life of Thomas Young, M.D., F.R.S.* London: John Murray, 1855, p. 348. Quoted in Robinson, Andrew. *The Last Man Who Knew Everything*. Oxford: One World, 2006, p. 191.

51 Clerke, Agnes Mary. 'Charles Babbage (1792–1871).' *Dictionary of National Biography (1885–1901)*. Ed. Gillian Fenwick. Oxford: Oxford University Press, 1989, p. 778.

52 Hall, Marie Boas. *All Scientists Now: The Royal Society in the Nineteenth Century*. Cambridge: Cambridge University Press, 1984, p. 30.

53 Hoskin, Michael. 'Astronomers at War: South V. Sheepshanks.' *Journal for the History of Astronomy* 20.62 (1989): 175–212, p. 190.

54 *Ibid.*, pp. 175, 203–4.

55 For full transcription see note 12, 'Calculation and Tabulation', Appendix I.

56 'This is the Engine which Charles Built', Royal Greenwich Observatory Archives (henceforth RGO). George Biddell Airy Papers, Cambridge RGO-452, Miscellaneous Mechanics, Hydraulics, f. 290.

57 Whewell to Herschel, 17 October 1822. Todhunter, I., ed. *William Whewell, D.D. Master of Trinity College, Cambridge: An Account of His Writings with Selections from His Literary and Scientific Correspondence*. Vol. 2. 2 vols. London: Macmillan, 1876, pp. 50–1.

58 *Autobiography*, p. 37.

59 Seelander, 13 November 1844. Quoted in Lindgren, Michael. *Glory and Failure: The Difference Engines of Johann Muller, Charles Babbage, and Georg and Edvard Scheutz*. Trans. Craig G McKay. 2 ed. Cambridge, MA: MIT Press, 1990, p. 133.

60 *Ibid.*, p. 197.

61 For a biographical account of Georg Scheutz see Lindgren (1990), pp. 82–97; for a brief biography of Edvard, *ibid.*, p. 116.

62 Airy, George Biddell. 'On Scheutz's Calculating Machine.' *Philosophical Magazine and Journal of Science* XII. July–December (1856): 225–6.

63 Wilkes, Maurice V. 'Babbage's Expectations for the Difference Engine.' *Annals of the History of Computing* 9.2 (1987), pp. 204–5. See Swade (2003), pp. 79–80.

64 L. J. Comrie revived the technique of verification by mechanical differencing using commercially available desktop calculators at the Nautical Almanac Office in the 1930s. The process was a major development in automated tabulation.

65 See Swade (2003), 256–7.

66 Trevelyan to Airy, 8 July 1857,T1/B/19264, PRO (Kew).

67 Airy to Hind, 18 August 1857. RGO6-454 f. 454 *et seq.*; Airy to Graham, 19 August 1857. RGO6-454 f. 458. For transcriptions of the letters see Swade (2003) Appendix II.

68 Hind to Airy, 27 August 1857. RGO6-454, f. 473.

69 Farr to Graham, 20 August 1857. RGO-454, f. 463. Swade (2003), p. 268.

70 Airy to Trevelyan, 30 September 1857. T1/6098B/19264.

71 *Ibid.*

72 For an account of Fowler's calculator see Swade (2000), pp. 310–12; Swade (2003), pp. 283–93; For reconstruction see Glusker, Mark, David M Hogan, and Pamela Vass. 'The Ternary Calculating Machine of Thomas Fowler.' *IEEE Annals of the History of Computing* 27.3 (2005): 4–22. Also: Glusker, M. 2005. *The ternary calculator of Thomas Fowler* [Online]. Available: http://mortati.com/glusker/fowler/.

73 Airy to Wheler, 19 May 1841, f. 56. Airy systematised the qualifications and level of competence of various grades of assistants at the Observatory. He formalised the requirements in arithmetic, mathematics, astronomy, written expression, and foreign language conversance for four grades of office: Supernumerary Computers, Junior Assistants, Superior Assistants, and Senior Assistant. The requirements were cumulative and increasingly demanding, each grade requiring the qualifications of the lower grade as well. See RGO6-814, ff. 197 (1–4). The grading sheets are dated 12 May 1857. They are listed in 'Printed Papers by G. B. Airy' (*Autobiography*, p. 384) under 'Knowledge expected in Computers and Assistants in the Royal Observatory'.

74 Bell to Airy, 8 August 1849, RGO6-428, f. 186.

75 Airy to Bell, 24 September 1849. RGO6-428, f. 190.

76 Jones, Matthew L. *Reckoning with Matter: Calculating Machines, Innovation and Thinking about Thinking from Pascal to Babbage.* Chicago, London: The University of Chicago Press, 2016, p. 13.

77 *Ibid.*, p. 65.

Chapter 4

1 Frank A. J. L. James, 'Christie, Samuel Hunter (1784–1865)', *Oxford Dictionary of National Biography*, Oxford University Press, 2004 [www.oxforddnb.com/view/article/5364, accessed 4 August 2018].

2 A. J. Meadows, 'Christie, Sir William Henry Mahoney (1845–1922)', *Oxford Dictionary of National Biography*, Oxford University Press, 2004 [www.oxforddnb.com/view/article/32409, accessed 4 August 2018].

3 'Sir William Christie, 1845–1922)' in *Proceedings of the Royal Society, Obituary Notices of Fellows Deceased*, 1922, pp. xi–xvi.

4 Meadows.

5 Emily Winterburn, *The Astronomers Royal* (Greenwich: National Maritime Museum, 2003); Meadows.

6 Royal Society obituary.

7 Winterburn.

8 Royal Society Obituary; 'Death of Sir W. Christie. A Great Astronomer', *Times*, 24 January 1922.

9 Winterburn.

10 Paul Gibbard, 'Bourdin, Martial (1867/8–1894)', *Oxford Dictionary of National Biography*, Oxford University Press, 2004 [www.oxforddnb.com/view/article/73217, accessed 4 August 2018]; Times reports, February–July 1894; Alex Butterworth, *The World that Never Was: A True Story of Dreamers, Schemers, Anarchists and Secret Agents* (London: The Bodley Head, 2010), pp. 329–334; ROG Learning Team, 'Propaganda by Deed: the Greenwich Observatory Bomb of 1894', Royal Museums Greenwich, 2005 [www.rmg.co.uk/explore/astronomy-and-time/astronomy-facts/history/propaganda-by-deed-the-greenwich-observatory-bomb-of-1894, accessed 25 December 2011, no longer available]; David Mulry, 'Popular Accounts of the Greenwich Bombing and Conrad's *The Secret Agent*', Rocky Mountain Review 43 (2000), 43–64, Wikipedia, 2007 [http://en.wikipedia.org/wiki/The_Secret_Agent, accessed 4 August 2018]; 'Theodore Kaczynski', Wikipedia, 2007 [http://en.wikipedia.org/wiki/Theodore_Kaczynski, accessed 4 August 2018].

11 E. Walter Maunder, *The Royal Observatory Greenwich: its History and Work* (London: The Religious Tract Society, 1900).

12 M. T. Brück, 'Maunder, Ernest Walter (1851–1928)', *Oxford Dictionary of National Biography*, Oxford University Press, 2004 [www.oxforddnb.com/view/article/40881, accessed 4 August 2018].

13 Wikipedia, 'Edward Walter Maunder' (2018) [https://en.wikipedia.org/wiki/Edward_Walter_Maunder, accessed 4 August 2018].

14 A. J. Meadows, 'Dyson, Sir Frank Watson (1868–1939)', *Oxford Dictionary of National Biography*, Oxford University Press, 2004 [www.oxforddnb.com/view/article/32949, accessed 4 August 2018]; A. S. Eddington, 'Sir Frank Watson Dyson 1868–1939', *Obituary Notices of Fellows of the Royal Society* 3 (1940), 159–172.

15 Margaret Wilson, *Ninth Astronomer Royal: The life of Frank Watson Dyson* (Cambridge: Heffer, 1951).

16 Eddington.

17 Obituary, Monthly Notices of the Royal Astronomical Society, 96 (February 1936), p. 297 [available at https://academic.oup.com/mnras/article/96/4/297/999336, accessed 4 August 2018]; H. P. Hollis, 'William Grasett Thackeray', The Observatory, February 1936, pp. 216–219 [available at http://adsabs.harvard.edu/full/1935Obs....58..216H, accessed 4 August 2018]. The RAS obituary shares some phrases with that in The Observatory and may also be by Hollis.

18 Winterburn; Meadows; Eddington; H. C. Plummer, 'Arthur Stanley Eddington, 1882–1944', *Obituary Notices of Fellows of the Royal Society* 5 (1945), pp. 113–125.

19 A. J. Meadows, 'Crommelin, Andrew Claude de la Cherois (1865–1939)', *Oxford Dictionary of National Biography*, Oxford University Press, 2007 [www.oxforddnb.com/view/article/57187, accessed 4 August 2018].

20 Richard Woolley, 'Charles Rundle Davidson 1875–1970', *Biographical Memoirs of Fellows of the Royal Society* 17 (1971), pp. 193–194.

21 David Ball, 'Cottingham, Edwin Turner (1869–1940) MODERN TIMES', Ringstead People (2010) [http://ringstead.squarespace.com/ringstead-people/2010/10/14/cottingham-edwin-turner-1869–1940-modern-times.html, accessed 4 August 2018].

22 C. M. Linton, *From Eudoxus to Einstein: A History of Mathematical Astronomy* (Cambridge: Cambridge University Press, 2004).

23 Eddington.

24 *The Times*, November 7, 1919.

25 Eddington.

26 Admiral of the Fleet the Earl of Cork and Orrery [W.H.D. Boyle], *My Naval Life* (London: Hutchinson, 1942).

27 Jonathan Betts, 'Gould, Rupert Thomas (1890–1948)', *Oxford Dictionary of National Biography*, Oxford University Press, 2013 [www.oxforddnb.com/view/article/40920, accessed 4 August 2018]; Jonathan Betts, *Time Restored: The Harrison Timekeepers and R.T. Gould, the Man Who Knew (Almost) Everything* (Oxford: Oxford University Press, 2006).

28 Wilson.

29 Edmund T. Whittaker, 'Philip Herbert Cowell 1870–1949', *Obituary Notices of Fellows of the Royal Society* 6 (1949), pp. 375–384. See also Graham Dolan, 'Philip Herbert Cowell' (2014) [www.royalobservatorygreenwich.org/articles.php?article=1199, accessed 4 August 2018].

30 Meadows, 'Crommelin'.

31 Harold Knox-Shaw (1885–1970) was Director of the Radcliffe Observatory, Oxford and subsequently of the Radcliffe Observatory, Pretoria.

32 Wilson.

33 Whittaker. The reference is to Robert Southey, 'The Battle of Blenheim'.

34 T. G. Cowling, 'Chapman, Sydney (1888–1970)', rev., *Oxford Dictionary of National Biography*, Oxford University Press, 2004 [www.oxforddnb.com/view/article/32367, accessed 4 August 2018]; T. G. Cowling, 'Sydney Chapman 1888–1970', *Biographical Memoirs of Fellows of the Royal Society* 17 (1971), pp. 53–89.

35 Harold Spencer Jones, 'John Jackson 1887–1958', *Biographical Memoirs of Fellows of the Royal Society* 5 (1960), pp. 95–106; George A. Wilkins, 'Jackson, John (1887–1958)', *Oxford Dictionary of National Biography*, Oxford University Press, 2004 [www.oxforddnb.com/view/article/57193, accessed 4 August 2018].

36 Hermann A. Brück, 'Greaves, William Michael Herbert (1897–1955)', *Oxford Dictionary of National Biography*, Oxford University Press, 2004 [www.oxforddnb.com/view/article/52569, accessed 4 August 2018]; R. O. Redman, 'William Michael Herbert Greaves 1897–1955', *Biographical Memoirs of Fellows of the Royal Society* 2 (1956), pp. 128–138.

37 R. V. D. R. Woolley, 'Harold Spencer Jones 1890–1960', *Biographical Memoirs of Fellows of the Royal Society* 7 (1961), pp. 136–145; D. H. Sadler, 'Jones, Sir Harold Spencer (1890–1960)', *Oxford Dictionary of National Biography*, Oxford University Press, 2004 [www.oxforddnb.com/view/article/34228, accessed 4 August 2018].

38 Winterburn.

39 Winterburn; Woolley; Chas Parker, 'Castle in the Sky – the Story of the Royal Greenwich Observatory at Herstmonceux', in Patrick Moore, ed., *The Yearbook of Astronomy 2000* (London: Macmillan, 2000).

40 William McCrea, 'Richard van der Riet Woolley 22 April 1906–24 December 1986', *Biographical Memoirs of Fellows of the Royal Society* 34 (1988), pp. 922–982;

William McCrea and Donald Lynden-Bell, revised by H. C. G. Matthew, 'Woolley, Sir Richard van der Riet (1906–1986)', *Oxford Dictionary of National Biography*, Oxford University Press, 2015 [www.oxforddnb.com/view/article/40030, accessed 4 August 2018].

41 McCrea.

42 'The Institute of Astronomy: Past and Present', Institute of Astronomy, University of Cambridge, 2003 [www.ast.cam.ac.uk/history, accessed 13 April 2007, no longer available], Wikipedia. 'Royal Observatory Greenwich' [https://en.wikipedia.org/wiki/Royal_Observatory_Greenwich, accessed 4 August 2018].

Chapter 5

1 Cooke John, Maule John, *An historical account of the Royal Hospital for Seamen at Greenwich*, London, 1789.

2 Masters and Mates were the seamen officers on all ships, Royal or Merchant. In the Royal Navy, Masters held warrants from the Navy Board that appointed them to particular ships, and in contrast to commission officers who often were put on half pay during times of peace, they remained on full pay. Masters were said to be of wardroom rank. The Mates, who also held warrants, were petty officers. There is an expression relating to Masters of ships, *Master after God*, which is highly descriptive of the powers and responsibilities of Masters in the old days. The Admiralty at one time noted rather sourly that the school's most intelligent boys opted for the Merchant Service. Masters in the Royal Navy received salaries between those of a lieutenant and post-captain, and when the intermediate rank of commander was introduced it actually was 'Master and Commander', until, once more, snobbishness defeated the old titles and ways of the sea as recorded in the Black Book of Admiralty, and in the Royal Navy, Master was dropped from the title, to leave the rank Commander. The title 'Master of the Fleet' for the Fleet Navigating Officer in a fleet such as the Home Fleet, or Mediterranean Fleet, existed until the 1950s.

3 John Seller, Hydrographer to the King, *Practical Navigation or An Introduction to the Whole Art* (London, Fourth edition, enlarged, 1680), 1.

4 See Doron Swade's chapter for a discussion of the elimination of errors in tabulated mathematical functions as a goal of Charles Babbage's difference engines. In fact the Greenwich Hospital School teacher Edward Riddle was well aware of Babbage's work, as they both sat, simultaneously on the committee of the Royal Astronomical Society at the time when Babbage announced that he had produced a table of logarithms of natural numbers for use of the Trigonometrical Survey of Ireland. At that time, Babbage also produced 'Notice respecting some Errors Common to many Tables of Logarithms'.

5 Mathematical table compilers often introduced a few deliberately small errors into their works in order to try and preserve their intellectual property rights.

6 The first specific mathematician teacher that can at present be identified as a teacher at GRHS is William Garrard, assistant to Astronomer Royal Maskelyne, and member of the Board of Longitude, who was appointed a 'Quarter Master' in 1808. In those days, in teaching, the title Master referred to the position that these days we would refer to as Head Teacher, so perhaps Quarter Master referred to something like head of department?

7 Hutton was born in 1737 in Newcastle-upon-Tyne. He never attended a university, and early decided that he would take on the role of teaching teachers. In 1773 he became Professor of Mathematics at the Royal Military Academy in Woolwich. In 1774 he became an FRS, and later he received an honorary doctorate from Edinburgh University.

8 His Notes, for example, indicate that he was aware of and understood Gauss's work on surfaces, and that he kept himself abreast of developments in his fields.

9 John William Norie (1772–1843), was a mathematician, chart maker and publisher of nautical books, in particular *Epitome of Practical Navigation* (1805), which became a standard work on navigation.

10 Hall & De Morgan to Admiralty, 15 Nov 1841, Nat Arch: ADM 7/61.

11 Education and Empire – Naval Tradition and England's Elite Schooling, D McLean, British Academic Press, 1999.

12 More information about Fisher can be found in Richard Dunn's chapter.

Chapter 6

1 Andrew Lambert, *John Knox Laughton, the Royal Navy and the Historical Profession* (London: Chatham, 1998), p. 33.

2 John Venn and J. A. Venn, *Alumni Cantabrigienses* (Cambridge: CUP, 1922–1954).

3 Anonymous, 'Carlton John Lambert', *Monthly Notices of the Royal Astronomical Society*, 83 (1922), pp. 244–245 [http://articles.adsabs.harvard.edu/full/seri/MNRAS/0083//0000244.000.html, accessed 5 August 2018].

4 Admiralty to President, RNC, 5 September 1877, National Archives ADM 203/2 and 21 August 1909, National Archives ADM203/10.

5 British Library, Jellicoe papers, Add. MS. 49038 (p. 139).

6 Jellicoe, *ibid.*, p. 134.

7 'Educational Staff and Salaries', undated (1873 or 1874), National Archives ADM 203/1.

8 Lambert, *John Knox Laughton*.

9 G. A. R. Callender, 'Laughton, Sir John Knox (1830–1915)', rev. Andrew Lambert, *Oxford Dictionary of National Biography*, Oxford University Press, 2007 [www.oxforddnb.com/view/article/34420, accessed 5 August 2018].

10 Letter from Laughton to the Admiral President, RNC, 31 March 1876, National Archives ADM 203/2.

11 Admiralty to President, RNC, 23 December 1884, National Archives ADM 203/4; Lambert, *John Knox Laughton*.

12 Admiralty to President, RNC, 24 September 1873, National Archives ADM 203/1.

13 W. H. Brock, 'Wormell, Richard (1838–1914)', *Oxford Dictionary of National Biography*, Oxford University Press, 2004 [www.oxforddnb.com/view/article/40970, accessed 5 August 2018].

14 Andrew Lambert (ed.), *Letters and Papers of Professor Sir John Knox Laughton, 1830–1915* (Aldershot: Ashgate for the Navy Records Society, 2002), pp. 15–19.

15 A Naval Nobody, 'On Naval Education', *MacMillan's Magazine* 37 (1878), 315–322.

16 Admiralty to President, RNC, 22 October 1878, National Archives ADM 203/2.

17 Admiralty to President, RNC, 30 June 1880, 11 March 1892 and 21 June 1890, National Archives ADM 203/3, and 203/5.

18 Admiralty to President, RNC, 20 March 1891, National Archives ADM 203/5.

19 E. A. Walker, 'Solomon, Sir Richard Prince (1850–1913)', rev. Christopher Saunders, *Oxford Dictionary of National Biography*, Oxford University Press, 2004; online edn, 2006 [www.oxforddnb.com/view/article/36184, accessed 5 August 2018].

20 Many letters Admiralty to President, Royal Naval College: National Archives ADM 203/11 and ADM 203/12.

21 Patrick Beesly, *Room 40: British Naval Intelligence 1914–1918* (London: Hamish Hamilton, 1982; Oxford: Oxford University Press, 1984).

22 Correspondence: National Archives T1/12294; G. H. Hardy, 'Mercer, James (1883–1932)', rev. John Bosnell, *Oxford Dictionary of National Biography*, Oxford University Press, 2004 [www.oxforddnb.com/view/article/34990, accessed 5 August 2018].

23 A. G. Howson, 'Charles Godfrey (1873–1924) and the Reform of Mathematical Education', *Educational Studies in Mathematics* 5 (1973), 157–180.

24 A. W. Siddons, 'Obituary: Charles Godfrey, M.V.O., M.A.', *The Mathematical Gazette* 12 (1924), 137–138.

25 *Ibid.*

26 For a fictional account of a public lecture by Cayley during this controversy, see Catherine Shaw, *The Three-Body Problem* (London: Allison and Busby, 2004), pp. 69–77.

27 Howson, 'Charles Godfrey'.

28 *Ibid.*

29 *Ibid.*

30 Beesly, *Room 40*.

31 Siddons, 'Charles Godfrey'.

32 Correspondence between Admiralty and Admiral President, Royal Naval College, May–December 1921, National Archives ADM 203/13.

33 Death notice, London Gazette, 2 July 1943; Venn, *Alumni Cantabrigiensis*. The letters concerning Witherington's appointment to the Professorship are absent from the archives and, unlike other RNC appointments, it was not announced in the Court pages of the Times. In the absence of any evidence to the contrary it seems likely that Witherington was appointed to the Chair to replace Godfrey in 1924 or 1925 and that it was following Witherington's retirement in 1933 or 1934 that Milne-Thomson was appointed Professor. I am grateful to Richard Rex and June Barrow-Green for their assistance in attempting to find further information about Witherington.

34 Admiralty to President, Royal Naval College, 26 July 1921, National Archives ADM 203/13.

35 President, Royal Naval College to Secretary to the Admiralty, 21 November 1923, National Archives ADM 203/14.

36 Mary Croarken, 'Thomson, Louis Melville Milne (1891–1974)', *Oxford Dictionary of National Biography*, Oxford University Press, 2004 [www.oxforddnb.com/view/article/51596, accessed 5 August 2018].

37 David Acheson, personal communication.

38 I am grateful to Dr Enid Michael for allowing me to read her father's unpublished autobiographical memoir.

39 The programme and accounts are in the BMC archive, held by the London Mathematical Society, but the list of participants does not appear to have survived. The accounts show 134 'registration fees' of 5s. and 5 of 10s, with three subsequently being refunded. Listed separately are 140 'fees' of 5s. and 15 of 10s., and 116 payments for meals are recorded. I am grateful to Miss Susan Oakes for her help in accessing this archive material.

40 J. W. S. Cassels, 'Louis Joel Mordell. 1888–1972', *Biographical Memoirs of Fellows of the Royal Society*, 19 (1973), pp. 493–520; Clive Kilmister (personal communication).

41 BMC minute-book, LMS archive.

42 Peter M. Neumann (personal communication).

43 Alan Camina (personal communication).

44 Clive Kilmister (personal communication).

45 E. A. Maxwell, 'Obituary: Thomas Arthur Alan Broadbent', *The Mathematical Gazette* 57 (1973), pp. 195–197; Clive Kilmister (personal communication); A. G. Howson (personal communication).

46 Howson (personal communication).

47 Kilmister (personal communication).

48 Howson (personal communication).

49 D. K. Brown, *A Century of Naval Construction: The History of the Royal Corps of Naval Constructors* (London: Conway, 1983).

50 John Bold, *Greenwich: An Architectural History of the Royal Hospital for Seamen and the Queen's House* (New Haven: Yale University Press, 2000).

51 McCrea, 'Woolley'.

52 June Barrow-Green (personal communication).

53 A. G. Howson (personal communication).

Chapter 7

1 For information about Thomas Hirst, see J. H. Gardner and R. J. Wilson, 'Thomas Archer Hirst – mathematician Xtravagant' (in six parts), in the *American Mathematical Monthly* 100 (1993): I. A Yorkshire surveyor, 435–441; II. Student days in Germany, 531–538; III. Göttingen and Berlin, 619–625; IV. Queenwood, France and Italy, 723–731; V. London in the 1860s, 827–834; VI. Years of decline, 905–913.

2 William Spottiswoode (1825–1883), mathematician and Treasurer of the Royal Society, was a close friend of Hirst and a fellow member of the X-Club (see later).

3 In 1873 Thomas Huxley (1825–1895) was Secretary of the Royal Society, a fellow member of the X-club, and advisor to the Committee of Council of Education, Kensington Museum, from where the letter was sent. The letter appears in Hirst's *Diaries*, pp. 1956–1957 (see Note 6).

4 The information about the Royal Naval College appears in their centenary booklet *Royal Naval College Greenwich Centenary 1873–1973*, published by the College.

5 This information was provided by the Public Records Office at Kew.

6 Of the 22 volumes of the diaries only one survives. A typescript of them all is housed in the library of the Royal Institution, London, and a facsimile reproduction has been published in microfiche: William H. Brock and Roy M. MacLeod (eds.), *Natural Knowledge in Social Context: The Journals of Thomas Archer Hirst FRS*, Mansell, London, 1980. It is referred to below as *Diaries*.

7 John Tyndall (1820–1893) was a distinguished popularizer of science. He was Hirst's closest friend, and another member of the X-club.

8 At the University of Marburg Hirst studied chemistry with Robert Bunsen (after whom the Bunsen burner is named), physics with Christian Gerling, and mathematics with Friedrich Stegmann.

9 Carl Friedrich Gauss (1777–1855); Wilhelm Eduard Weber (1804–1891); Johann Peter Gustav Lejeune-Dirichlet (1805–1859); Jacob Steiner (1796–1863). As a pure geometer with similar interests, Steiner had a great influence on Hirst. See also Robin Wilson, 'What was Jakob Steiner like?' and 'What was Lejeune Dirichlet like?', *Mathematical Intelligencer* 40 (3), 2018, 51–56, and 41 (1), 2019, 51–55.

10 Joseph Liouville (1809–1882); Michel Chasles (1793–1880); Joseph-Louis Bertrand (1824–1900); Francesco Brioschi (1824–1897); Antonio Luigi Gaudenzio Giuseppe Cremona (1830–1903).

11 Arthur Cayley (1821–1895); James Joseph Sylvester (1814–1897); George Boole (1815–1864).

12 Pafnuty Lvovich Chebyshev, or Tchebichef (1821–1894).

13 The nine members all acquired nicknames based on X, such as the X̲alted Huxley, the X̲centric Tyndall, the X̲cellent Spottiswoode, and the X̲travagant Hirst.

14 Augustus De Morgan (1806–1871), Professor of Mathematics at University College, London.

15 In the 1880s the Association for the Improvement of Geometrical Teaching widened its scope to other areas of mathematics. It still exists under its later name of the Mathematical Association.

16 The diary extracts in this section appear in *Diaries*, pp. 1956, 1958–1959, 1965, and 1967.

17 George Joachim Goschen, (1831–1907), Liberal Member of Parliament, was Chancellor of the Duchy of Lancaster and later Chancellor of the Exchequer. In 1900 he became First Viscount Goschen.

18 Astley Cooper Key (1821–1888) was knighted and appointed Vice-Admiral on his appointment as President of the Royal Naval College. He later became a full Admiral.

19 The 'Scientific Person' was Ellen Lubbock (1834/5–1879), Hirst's 'dearest, truest and most constant lady friend'. Her husband Sir John Lubbock, archaeologist, was another member of the X-club and became the first Lord Avebury. In the first line of the verse 'T.G.!' should be 'G.G.!' (George Goschen).

20 The diary extracts in this section appear in *Diaries*, pp. 1961, 1963, 1970, 1979, and 1984.

21 Arnold William Reinold (1843–1921) served as Professor of Physics at the Royal Naval College from 1873 to 1908; he was appointed a Fellow of the Royal Society in 1883. Heinrich Debus (1824–1916) was Professor of Chemistry from 1873 to 1888, jointly with an appointment at Guy's Hospital; Kalley Miller (1843–1889) remained at the College from 1873 to 1885.

22 This information was provided by the Public Records Office at Kew.

23 The Shah of Persia visited London from 19 June to 5 July 1873.

24 The diary extracts in this section appear in *Diaries*, pp. 1984, 1993, 1999, 2006, 2011, 2019–2020, 2022–2023, 2025–2026, 2035, 2082, 2084, and 2112. Christian Felix Klein (1849–1925).

25 Sir George Biddell Airy (1801–1892) was appointed Astronomer Royal in 1835.

26 William Henry Smith (1825–1891), bookseller, later First Lord of the Treasury.

27 The geometry publications that Hirst completed while at Greenwich are:
On the correlation of two planes, *Proc. L.M.S.* 5 (1873/4), 40–70.
On correlation in space, *Proc. L.M.S.* 6 (1874/5), 7–9.
Notes on the correlation of two planes, *Proc. L.M.S.* 8 (1876/7), 262–273.

Note on the complexes generated by two correlative planes, *Proc. L.M.S.* 10 (1878/9), 131–153.

On the complexes generated by two correlated planes, in *Collectanea Mathematica*, in memoria Domenici Chelini (ed. L. Cremona and E. Beltrami), Mediolini, Naples, 1881, pp. 51–73.

On quadric transformation, *Quart. J. Math.* 17 (1881) 301–311.

On Cremonian congruences, *Proc. L.M.S.* 14 (1882/3), 259–301.

He also compiled a booklet, *Geometrical Contributions to the 'Educational Times'*, published by Hodgson, London, in 1875.

28 The diary extracts in this section appear in *Diaries*, pp. 1979, 1999, 2005, 2010, 2012, 2017, 2021–2023, 2075, 2106, 2114, and 2132–2133.

29 Friedrich Otto Rudolf Sturm (1841–1891).

30 This appears in a letter from Smith to Hirst dated 23 September 1874, in the *Hirst Correspondence* at University College, London.

31 Marie Ennemond Camille Jordan (1838–1922); for Chasles, Bertrand and Tchebichef, see Notes 10 and 12.

32 The diary extracts in this section can be found in *Diaries*, pp. 1993–1994, 1996, 2001, 2026, 2045, 2056, 2116, and 2122.

33 Sir Joseph Lister (1827–1912), later the first Baron Lister and President of the Royal Society.

34 The diary extracts in this section can be found in *Diaries*, pp. 2128 and 2130–2132.

35 William Davidson Niven (1842–1917), from Trinity College, Cambridge, was later knighted.

Chapter 8

1 A. R. Forsyth, 'William Burnside', *Nature*, 120 (1927), 555–557; *Proc. Roy. Soc.*, Series A, 117 (1928), *xi–xxx=J. London Math. Soc.*, 3 (1928), 64–80=Preface to *Theory of Probability* [W. Burnside, *Theory of Probability* (edited by A. R. Forsyth), CUP, Cambridge 1928 (reprinted by Dover, New York 1959)]; reprinted in *The Collected Papers of William Burnside* (2 volumes edited by Peter M. Neumann, A. J. S. Mann and Julia C. Tompson), OUP, Oxford, 2004, ISBN 0-19-850585-X, I, pp. 67–84.

2 Martin G. Everett, A. J. S. Mann and Kathy Young, 'Notes on Burnside's life', *The Collected Papers*, I, pp. 85–109.

3 Thomas Hawkins, 'William Burnside', *Dictionary of Scientific Biography*, Vol. 2, Scribner, New York, 1971, p.615.

4 *The Collected Papers*; *Theory of Groups of Finite Order*, CUP, Cambridge 1897. Second edition, CUP, Cambridge, 1911 (reprinted by Dover, New York 1955); *Theory of Probability* (edited by A. R. Forsyth), CUP, Cambridge, 1928 (reprinted by Dover, New York 1959).

5 Peter M. Neumann, 'The context of Burnside's contributions to group theory', *The Collected Papers* I, 15–37.

6 *Theory of Probability*.

7 *The Collected Papers*, I.

8 Camille Jordan, *Traité des substitutions et des équations algébriques*, Gauthier-Villars, Paris 1870.

9 Walter Feit and John G. Thompson, 'Solvability of Groups of Odd Order', *Pacific J. Math.*, 13 (1963), 775–1029.

10 *Theory of Groups of Finite Order* (first edition).
11 W. Burnside, 'On the theory of groups of finite order (Presidential Address)', *The Collected Papers, II*, 1235–1241.

Chapter 9

1 'Report of the Committee of the Astronomical Society of London relative to the Improvement of the Nautical Almanac' *Memoirs of the Astronomical Society*, 1831, vol. 4, pp. 447–470.
J.L.E. Dreyer and H.H. Turner, *History of the Royal Astronomical Society 1820–1920*, London: Royal Astronomical Society, 1923, pp. 56–63.
W.J. Ashworth, 'The Calculating Eye: Baily, Herschel, Babbage and the Business of Astronomy', *British Journal for the History of Science*, 1994, vol. 27, pp. 409–441.
2 D.H. Sadler, 'A personal history of H.M. Nautical Almanac Office 30 October 1930–18 February 1972', Ed. G.A. Wilkins, unpublished typescript. Copy in author's possession. Another copy is deposited in the Royal Greenwich Observatory Archives held in Cambridge University Library.
3 H.S.W. Massey, 'L.J. Comrie', *Obituary Notices of Fellows of the Royal Society*, 1953, vol. 8, pp. 97–101, 100.
4 M. Croarken, 'Tabulating the Heavens: Computing the Nautical Almanac in 18th-Century England', *IEEE Annals of the History of Computing*, 2003, vol. 25, no. 3, pp. 48–61.
5 T.C. Hansard, Parliamentary Debates, Vol. XXXVII, March 6th 1818, 877–878.
6 D.H. Sadler, 'The Nautical Almanac in its Seventh Third of a Century', *Navigation*, Winter 1967, vol. 14, no. 4, pp. 349–355.
E.G. Forbes, 'The Foundation and Early Development of the Nautical Almanac', *Journal of the Institute of Navigation*, 1965, vol. 18, no. 4, pp. 391–401.
Act of Parliament 1818 'for more effectually discovering Longitude at sea...' 58 George III, cap. XX.
7 'Report of the Committee of the Astronomical Society of London relative to the Improvement of the Nautical Almanac', *Memoirs of the Astronomical Society*, 1831, vol. 4, pp. 447–470.
8 For brief histories of the Nautical Almanac Office see
G.A. Wilkins, 'The History of H.M. Nautical Almanac Office', pp. 54–81 in A.D. Fiala and S.J. Dick (eds.), *Proceedings Nautical Almanac Office Sesquicentennial Symposium US Naval Observatory March 3–4 1999*, Washington, DC: US Naval Observatory.
G.A. Wilkins, 'The Making of Astronomical Tables in HM Nautical Almanac Office', pp. 294–320 in M. Campbell-Kelly, M. Croarken, R. Flood, and E. Robson (eds.), *The History of Mathematical Tables: From Sumer to Spreadsheets*, Oxford: Oxford University Press, 2003.
9 Cambridge University Library Manuscripts RGO 16/7 Packet 11.
10 E.T. Whittaker, 'Philip Herbert Cowell', *Obituary Notices of Fellows of the Royal Society*, 1948–1949, vol. 6, pp. 375–384.
11 W.B. Lord, 'The "Nautical Almanac" Office', *Nautical Magazine*, 1897, vol. 66, pp. 435–446.

12 W.H.M. Greaves, 'Obituary Notices: Leslie John Comrie', *Monthly Notices of the Royal Astronomical Society*, 1953, vol. 113, no. 3, pp. 294–304.
M. Croarken, *Early Scientific Computing in Britain,* Oxford: Clarendon Press, 1990.
Cambridge University Library Manuscripts RGO 16/7 Packet 9 entry 12.12.1910.

13 T.C. Hudson, 'H.M. Almanac Office Anti-Differencing Machine', pp. 127–131, in E.M. Horsburgh (ed.), Napier Tercentenary Celebration *Handbook of the Exhibition of Napier Relics and of Books, Instruments and Devices for Facilitating Calculation,* Edinburgh: Royal Society of Edinburgh, 1914.

14 G.A. Wilkins, 'The Making of Astronomical Tables in HM Nautical Almanac Office', pp. 294–320, in M. Campbell-Kelly, M. Croarken, R. Flood, and E. Robson (eds.), *The History of Mathematical Tables: From Sumer to Spreadsheets*, Oxford: Oxford University Press, 2003.

15 L.J. Comrie, 'Careers for Girls', *Mathematical Gazette*, 1944, vol. 28, pp. 90–95.

16 L.J. Comrie, 'First Report of the Computing Section', *Memoirs of the British Astronomical Association*, vol. 24, Reports of the Sections pp. 1–28.

17 For example, L.J. Comrie, 'The Application of Calculating Machines to Astronomical Computing', *Popular Astronomy*, 1925, vol. 33, pp. 243–246.

18 E.T. Whittaker, 'Philip Herbert Cowell', *Obituary Notices of Fellows of the Royal Society*, 1948–1949, vol. 6, pp. 375–384.

19 Cambridge University Library Manuscripts RGO 16/7 Packet 12 Letter from P.H. Cowell to Secretary of the Admiralty 28 August 1928.

20 M. Croarken, *Early Scientific Computing in Britain*, Oxford: Clarendon Press, 1990, pp. 40–41.
D.H. Sadler, 'A Personal History of H.M. Nautical Almanac Office 30 October 1930–18 February 1972', Ed. G.A. Wilkins, unpublished typescript, p. 29.

21 L.J. Comrie, 'Nautical Almanac Office: Description of Tables in Preparation. Methods of Preparation. Progress of Work', 1926 July 24, typescript, Cambridge University Library, RGO 16/1 Misc 8.

22 D.H. Sadler, 'A Personal History of H.M. Nautical Almanac Office 30 October 1930–18 February 1972', Ed. G.A. Wilkins, unpublished typescript, chapter 2.
Exchange of letters between E. Sawer and L.J. Comrie 1931, Cambridge University Library Manuscripts RG) 16/5 Packet 5.
L.J. Comrie, 'Nautical Almanac Office: Description of Tables in Preparation. Methods of Preparation. Progress of Work', 1926 July 24, typescript, Cambridge University Library, RGO 16/1 Misc 8.
Cambridge University Library, RGO 16/17 War Office Packet Letter L.J. Comrie to Colonel Winterbotham 20 January 1928.

23 L.J. Comrie, 'On the construction of tables by interpolation', *Monthly Notices of the Royal Astronomical Society*, 1928, vol. 88, pp. 506–523.

24 L.J. Comrie, 'On the construction of tables by interpolation', *Monthly Notices of the Royal Astronomical Society*, 1928, vol. 88, pp. 506–523.

25 L.J. Comrie, *Interpolation and Allied Tables*, London: HMSO, 1936.

26 L.J. Comrie, 'On the application of the Brunsviga-Dupla calculating machine to double summation with finite differences', *Monthly Notices of the Royal Astronomical Society*, 1928, vol. 88, no. 5, pp. 447–459.

27 M. Croarken, *Early Scientific Computing in Britain*, Oxford: Clarendon Press, 1990, pp. 27–30.
L.J. Comrie, 'The Nautical Almanac Office Burroughs Machine', *Monthly Notices of the Royal Astronomical Society*, 1932, vol. 92, no. 6, pp. 523–541.

28 E.W. Brown, *Tables of the Motion of the Moon*, 3 vols, New Haven: Yale University Press, 1919.

29 L.J. Comrie, 'The Application of the Hollerith Tabulating Machine to Brown's Tables of the Moon', *Monthly Notices of the Royal Astronomical Society*, 1932, vol. 92, pp. 694–707.

30 Cambridge University Library, RGO 16/7 Packet 12, P.H. Cowell to Secretary of the Admiralty 2 May 1928, Admiralty Secretary to P.H. Cowell 10 December 1928.

31 D.A. Grier, *When Computers Were Human*, Princeton: Princeton University Press, 2005.

   M. Croarken, *Early Scientific Computing in Britain*, Oxford: Clarendon Press, 1990.

32 D.H. Sadler, 'A Personal History of H.M. Nautical Almanac Office 30 October 1930–18 February 1972', Ed. G.A. Wilkins, unpublished typescript, pp. 20–21.

33 L.J. Comrie, 'Modern Babbage Machines', *Bulletin of the Office Machinery User's Association 1931–1932*, 29pp.

   L.J. Comrie, 'Inverse Interpolation and Scientific Applications of the National Accounting Machine', *Supplement to the Royal Statistical Society*, 1936, vol. 3, no. 2, pp. 87–114.

   M. Croarken, *Early Scientific Computing in Britain*, Oxford: Clarendon Press, 1990, pp. 33–36.

   M.R. Williams, 'Difference Engines: from Müller to Comrie', pp. 123–142 in M. Campbell-Kelly, M. Croarken, R. Flood, and E. Robson, *The History of Mathematical Tables: From Sumer to Spreadsheets*, Oxford: Oxford University Press, 2003.

34 Cambridge University Library, RGO 16/7 Packet 12 Letter L.J. Comrie to Secretary Admiralty 29 January 1934.

   D.H. Sadler, 'A personal history of H.M. Nautical Almanac Office 30 October 1930–18 February 1972', Ed. G.A. Wilkins, unpublished typescript, p. 27.

35 D.H. Sadler, 'A personal history of H.M. Nautical Almanac Office 30 October 1930–18 February 1972', Ed. G.A. Wilkins, unpublished typescript, p. 23.

36 Cambridge University Library, RGO 16/7 Packet 12 Letter from P.H. Cowell to Secretary Admiralty 28 August 1928.

37 L.J. Comrie, *Barlow's Tables*, 3rd edition, London: Spon, 1930.

38 L.J. Comrie and L.M. Milne-Thomson, *Standard Four-Figure Logarithms*, 2 vols, London: Macmillan, 1931.

39 R.C. Archibald, *Mathematical Table Makers: Scripta Mathematica Studies Number 3*, New York: Yeshiva University, 1948.

40 W.H.M. Greaves, 'Obituary Notices: Leslie John Comrie', *Monthly Notices of the Royal Astronomical Society*, 1953, vol. 113, no. 3, pp. 294–304.

41 J. Peters and L.J. Comrie, *Achtstellige Tafel der trigonometrischen Funktionen für jede Sexagesimalsekunde des Quadranten*, Berlin: Landesaufnahme, 1939.

42 M. Croarken and M. Campbell-Kelly, 'Beautiful Numbers: The Rise and Decline of the British Association Mathematical Tables Committee, 1871–1965', *IEEE Annals of the History of Computing*, 2000, vol. 22, no. 4, pp. 44–61.

   M. Croarken, 'Table Making by Committee: British Table Makers 1871–1965', pp. 234–263 in M. Campbell-Kelly, M. Croarken, R. Flood, and E. Robson, *The History of Mathematical Tables: From Sumer to Spreadsheets*, Oxford: Oxford University Press, 2003.

43 D.H. Sadler, 'A personal history of H.M. Nautical Almanac Office 30 October 1930–18 February 1972', Ed. G.A. Wilkins, unpublished typescript, p. 17.

44 Cambridge University Library Manuscripts RGO 16/15.
45 M. Croarken, *Early Scientific Computing in Britain*, Oxford: Clarendon Press, 1990, p. 41.
46 L.J. Comrie, *Tables of tan⁻¹x and log (1 + x²) to Assist in the Calculation of the Ordinates of the Pearson Type IV Curve*, Tracts for Computers, No. 23. London: Cambridge University Press, 1938.
47 G.B. Hey, Interview with M. Croarken, 22 November 1983.
    L.J. Comrie, G.B. Hey and H.B. Hudson, 'The Application of Hollerith Equipment to an Agricultural Investigation', *Supplement to the Journal of the Royal Statistical Society*, 1937, vol. 4, no. 2, pp. 210–224.
48 UK National Archives ADM 178/166, 18 August 1934.
49 M. Croarken, 'Case 5,656: L.J. Comrie and the Origins of the Scientific Computing Service Ltd', *IEEE Annals of the History of Computing*, 1999, vol. 21, no. 4, pp. 70–71.
50 Cambridge University Library Manuscripts RGO 16/17 Packet Royal Naval College.
51 UK National Archives ADM 178/166, ADM 178/167 and ADM 178/168.
52 M. Croarken, *Early Scientific Computing in Britain*, Oxford: Clarendon Press, 1990.
53 L.J. Comrie, *Chambers's Six-Figure Mathematical Tables. Vol. I Logarithmic Values. Vol II Natural Values*. London: Chambers, 1948.
    A. Fletcher, L. Rosenhead, J.C.P. Miller and L.J. Comrie, *An Index of Mathematical Tables Vol. 1; An Index of Mathematical Tables. Vol. II*, 2nd edition. (1st 1946), Oxford: Blackwell for Scientific Computing Service Ltd., 1962.

Chapter 10

1 Kevin Littlewood and Beverley Butler, *Of Ships and Stars. Maritime Heritage and the Founding of the National Maritime Museum, Greenwich*, Athlone Press, London & New Brunswick, 1998. The Museum was rebranded as Royal Museums Greenwich in 2012, but its legal title is still National Maritime Museum. This is the name used in this chapter.
2 The term 'scientific instrument' began to appear only in the nineteenth century. Deborah Warner, 'What is a Scientific Instrument, When Did it Become One, and Why?', *British Journal for the History of Science*, 23 (1990), pp. 83–93; Liba Taub, 'On Scientific Instruments', *Studies in History and Philosophy of Science*, 40 (2009), pp. 337–343, 337–338.
3 John Millburn, *Adams of Fleet Street, Instrument Makers to King George III*, Ashgate, Aldershot, 2000, pp. 362–382, illustrates the range offered by one retailer.
4 Littlewood and Butler, *Of Ships and Stars*, p. 55.
5 Maria Blyzinsky, 'The History of the Collection', in Elly Dekker, *Globes at Greenwich. A Catalogue of the Globes and Armillary Spheres in the National Maritime Museum, Greenwich*, Oxford University Press/National Maritime Museum, Oxford, 1999, pp. 13–20; Richard Dunn, 'Sundials and the Arts', in Hester Higton, *Sundials at Greenwich. A Catalogue of the Sundials, Horary Quadrants and Nocturnals in the National Maritime Museum, Greenwich*, Oxford University Press/National Maritime Museum, Oxford, 2002, pp. 39–49; Richard Dunn, 'Collecting and Interpreting Navigation at Greenwich', in Willem Mörzer Bruyns, *Sextants at Greenwich. A Catalogue of the Mariner's Quadrants, Mariner's Astrolabes, Cross-staffs, Backstaffs, Octants, Sextants, Quintants, Reflecting*

Circles and Artificial Horizons in the National Maritime Museum, Greenwich, Oxford University Press/NMM, Oxford, 2009, pp. 71–82; Anita McConnell, 'Fisher, George (1794–1873)', *Oxford Dictionary of National Biography*, Oxford University Press, 2004; online edn, Sept 2014, www.oxforddnb.com/view/article/9495, accessed 18 Oct 2017; George Roberts, 'Magnetism and Chronometers: the Research of the Reverend George Fisher', *British Journal for the History of Science*, 42 (2009), pp. 57–72.

6 The transfer of the Royal Observatory buildings took place between 1953 and 1958.

7 Koenraad van Cleempoel, *Astrolabes at Greenwich. A Catalogue of the Astrolabes in the National Maritime Museum*, Oxford University Press/National Maritime Museum, Oxford, 2006.

8 Mörzer Bruyns, *Sextants at Greenwich.*

9 Jonathan Betts, *Marine Chronometers at Greenwich*, Oxford University Press/ Royal Museums Greenwich, Oxford, 2018.

10 Higton, *Sundials at Greenwich.*

11 Warrant from Charles II to Sir Thomas Cricheley, Master-General of the Ordnance, quoted in Derek Howse, *Greenwich Time and the Longitude*, Philip Wilson/ National Maritime Museum, London, 1997, p. 44.

12 John Flamsteed, *Historia Coelestis Britannica*, London, 1725.

13 Quoted in Richard Sorrenson, 'George Graham, Visible Technician', *British Journal for the History of Science*, 32 (1999), pp. 202–221, 212; see also Richard Dunn, 'A Bird in the Hand, or, Manufacturing Credibility in the Instruments of Enlightenment Science', in A. Craciun and S. Schaffer (eds.), *The Material Cultures of Enlightenment Arts and Sciences* (Palgrave Macmillan, 2016), pp. 73–90, 76–79; Jeremy Lancelotte Evans, 'Graham, George (c.1673–1751)', *Oxford Dictionary of National Biography*, Oxford University Press, 2004; online edn, May 2005, www.oxforddnb.com/view/article/11190, accessed 12 Oct 2017.

14 In later years, however, the weight of the iron frame caused the instrument to deform.

15 William Ludlam, *An Introduction and Notes on Mr. Bird's Method of Dividing Astronomical Instruments*, London, 1786, p. iii, quoted in Sorrenson, 'George Graham', p. 219.

16 J.A. Bennett, 'The English Quadrant in Europe: Instruments and the Growth of Consensus in Practical Astronomy', *Journal for the History of Astronomy*, 23 (1992), pp. 1–14; Derek Howse, *Greenwich Observatory, Volume 3 The Buildings and Instruments*, Taylor & Francis, London, 1975, pp. 21–24; Alison Morrison-Low, *Making Scientific Instruments in the Industrial Revolution*, Ashgate, Aldershot, 2007, pp. 136–137; Sorrenson, 'George Graham', pp. 219–220.

17 Gilbert Satterthwaite, 'Airy's Transit Circle', *Journal of Astronomical History and Heritage*, 4 (2001), pp. 115–141.

18 Howse, *Greenwich Observatory*, pp. 43–48; Howse, *Greenwich Time*, pp. 133–143; Charles W. J. Withers, *Zero Degrees: Geographies of the Prime Meridian*, Harvard University Press, Cambridge MA, 2017.

19 Richard Dunn and Rebekah Higgitt, *Finding Longitude: How Ships, Clocks and Stars Helped Solve the Longitude Problem*, Collins, Glasgow, 2014, pp. 36–39; Derek Howse, 'Britain's Board of Longitude: The Finances, 1714–1828', *The Mariner's Mirror*, 84 (1998), pp. 400–417; Peter Johnson, 'The Board of Longitude 1714–1828', *Journal of the British Astronomical Association*, 99 (1989), pp. 63–69; Anthony Turner, 'Longitude Finding', in John B. Hattendorf (ed.), *The Oxford Encyclopedia of Maritime History*, Oxford University Press, 2007, pp. 405–415.

20 William Andrewes, *The Quest for Longitude*, Collection of Historical Scientific Instruments, Harvard University, Cambridge, Mass., 1996; Dunn and Higgitt, *Finding Longitude*; Rupert Gould, *The Marine Chronometer*, J.D. Potter, London, 1923; Dava Sobel, *Longitude*, Fourth Estate, London, 1996.

21 Howse, *Greenwich Time*, pp. 183–185; George Huxtable, 'Finding Longitude by Lunar Distance', *Navigation News*, September/October 2007, pp. 22–23.

22 Jim Bennett, 'Catadioptrics and Commerce in Eighteenth-century London', *History of Science*, 44 (2006), pp. 246–278, discusses the longer history and commercial aspects of this story. See also Silvio Bedini, *Thinkers and Tinkers. Early American Men of Science*, Scribner, New York, 1975, pp. 118–123; Dunn and Higgitt, *Finding Longitude*, pp. 93–95; Peter Ifland, *Taking the Stars. Celestial Navigation from Argonauts to Astronauts*, Krieger Publishing Co., Malabar, 1998, pp. 15–18; Mörzer Bruyns, *Sextants at Greenwich*, pp. 23–29.

23 Alan Stimson, 'The Influence of the Royal Observatory at Greenwich upon the Design of 17th and 18th Century Angle-measuring Instruments at Sea', *Vistas in Astronomy*, 20 (1976), pp. 123–130, 126.

24 John Hadley, 'An Account of Observations Made on Board the Chatham-Yacht, August 30th and 31st, and September 1st, 1732', *Philosophical Transactions*, 37 (1731–1732), pp. 341–356.

25 The quadrant is on display in the Observatory next to Graham's mural quadrant, the scale of which Bird re-divided in 1753.

26 Allan Chapman, 'Astronomia Practica: The Principal Instruments and their Uses at the Royal Observatory', *Vistas in Astronomy*, 20 (1976), pp. 141–156, 151–152. Bradley also had a productive relationship with George Graham, whose work he acknowledged; see Sorrenson, 'George Graham'; Dunn, 'A Bird in the Hand', pp. 79–81.

27 Board of Longitude, confirmed minutes, 6 March 1756, Cambridge University Library RGO 14/5, p. 21; Eric G. Forbes, *Tobias Mayer (1723–1762), Pioneer of Enlightened Science in Germany*, Vandenhoeck & Ruprecht, Göttingen, 1980, pp. 162–169.

28 John Charnock, *Biographia navalis*, London, 1794, p. 36; Derek Howse, 'Campbell, John (b. in or before 1720, d. 1790)', *Oxford Dictionary of National Biography*, Oxford University Press, 2004; online edn, Jan 2011, www.oxforddnb.com/view/article/4518, accessed 12 Oct 2017.

29 Chapman, 'Astronomia Practica', pp. 148–152; Dunn and Higgitt, *Finding Longitude*, pp. 95–99; Mörzer Bruyns, *Sextants at Greenwich*, p. 37; Dunn, 'A Bird in the Hand', pp. 83–86; Saul Moskowitz, 'The World's First Sextants', *Journal of The Institute of Navigation*, 34 (1987), pp. 22–42; Alan Stimson, 'Some Board of Longitude Instruments in the Nineteenth Century', in Peter de Clercq (ed.), *Nineteenth-Century Scientific Instruments and their Makers*, Amsterdam & Leiden, 1985, pp. 93–115, 93–98; Deborah Warner, 'John Bird and the Origin of the Sextant', *Rittenhouse*, 12 (1988–9), pp. 1–11.

30 Charles H. Cotter, *A History of Nautical Astronomy*, Hollis & Carter, London, 1968, pp. 195–242; Mary Croarken, 'Tabulating the Heavens: Computing the *Nautical Almanac* in 18th-century England', *IEEE Annals of the History of Computing*, 25 (2003), pp. 48–61; Mary Croarken, 'Nevil Maskelyne and His Human Computers', in Rebekah Higgitt (ed.), *Maskelyne: Astronomer Royal*, Hale Books, London, 2014, pp. 130–161; Dunn and Higgitt, *Finding Longitude*, pp. 104–114; Howse, *Greenwich Time*, pp. 62–71; Derek Howse, *Nevil Maskelyne*, Cambridge University

Press, Cambridge, 1989, pp. 59, 85–96; D.H. Sadler, *Man is Not Lost: A Record of Two Hundred Years of Astronomical Navigation with The Nautical Almanac 1767–1967*, Royal Greenwich Observatory, London, 1967.

31 In systems of equal hours, the day is divided into 24 h each of equal length. Systems using unequal hours divide the period of daylight into twelve equal parts and the period of night into twelve parts. The lengths of the hours therefore vary throughout the year, with day and night hours being of equal length only at the equinoxes.

32 Higton, *Sundials at Greenwich*, p. 337.

33 David J. Bryden, 'The Instrument-Maker and the Printer: Paper Instruments Made in Seventeenth Century London', *Bulletin of the Scientific Instrument Society*, 55 (1997), pp. 3–15; Boris Jardine, 'The "Incomparable" Mr. Sutton: A Famous 17th-century Instrument Maker', *Explore Whipple Collections*, Whipple Museum of the History of Science, University of Cambridge, 2006, www.hps.cam.ac.uk/whipple/explore/astronomy/theincomparablemrsutton/, accessed 12 June 2006.

34 Henry Sutton, *Description & Use of a Large Quadrant, Contrived and Made by H. Sutton : Accomodated with Various Lines, for the Easie Resolving of all Astronomical, Geometrical, and Gnomonical Problems, for Working of Proportions, and for Finding the Hour Universally*, printed by W. Godbid for R. Morden, London, 1669. See also Boris Jardine, 'The Sutton-type quadrant', *Explore Whipple Collections*, Whipple Museum of the History of Science, University of Cambridge, 2006, www.hps.cam.ac.uk/whipple/explore/astronomy/theincomparablemrsutton/suttonsquadrant/, accessed 12 June 2006.

35 Edmund Stone, 'Advertisement', in Nicolas Bion, *The Construction and Principal Uses of Mathematical Instruments*, translated and enlarged by Edmund Stone, Printed for J. Richardson, London, 1758; Jardine, 'The "Incomparable" Mr. Sutton'.

36 24 Geo 2 c. 23, 'An Act for regulating the Commencement of the Year, and for correcting the Calendar now in use', www.legislation.gov.uk/apgb/Geo2/24/23/introduction, accessed 12 Oct 2017; see also Duncan Steel, *Marking Time. The Epic Quest to Invent the Perfect Calendar*, John Wiley & Sons, New York, 2000, p. 240.

37 Quoted in Robert Poole, *Time's Alteration. Calendar Reform in Early Modern England*, UCL Press, London, 1998, p. 118.

38 Poole, *Time's Alteration*; Steel, *Marking Time*, pp. 233–250.

39 NMM NAV1042 is an example of the same quadrant with an 'old style' declination scale.

40 Bryden, 'The Instrument-Maker and the Printer', pp. 4–5, 13. For a more detailed description of the earlier version of the quadrant, see Higton, *Sundials at Greenwich*, pp. 348–351.

41 Poole, *Time's Alteration*, Ch. 1.

42 J.P. Rigaud, *Miscellaneous Works and Correspondence of James Bradley*, Oxford, 1832, pp. lxxx–lxxxii, quoted in Poole, *Time's Alteration*, pp. 6–7.

43 D.J. Bryden, *Napier's Bones. A History and Instruction Manual*, Harriet Wynter Ltd, London, 1992.

44 *Ibid.*, p. 14.

45 Moore and Pepys maintained their friendship and in 1673 were both involved in the foundation of the 'Royal Mathematical School within Christ's Hospital', which was to provide training in navigation for boys in the King's service at sea.

46 Howse, *Greenwich Time*, pp. 34–48; Frances Willmoth, 'Moore, Sir Jonas (1617–1679)', *Oxford Dictionary of National Biography*, Oxford University Press, 2004; online edn, April 2016, www.oxforddnb.com/view/article/19137, accessed 12 Oct 2017.

47 Samuel Butler, *Hudibras. The Second Part*, London, 1664, p. 209.
48 W.H. Quarrell and M. Mare, *London in 1710. From the Travels of Zacharias Conrad von Uffenbach*, Faber & Faber, London, 1934, p. 20.
49 Bryden, *Napier's Bones*, p. 24.
50 Quoted in Richard Larn, *Goodwin Sands Shipwrecks*, David & Charles, Newton Abbot & London, 1981, 1977, p. 53.
51 Martin Brayne, *The Greatest Storm*, Sutton Publishing, Stroud, 2002; L.G. Carr Laughton and V. Heddon, *Great Storms*, Philip Allan and Co. Ltd, London, 1930, pp. 68–98; Larn, *Goodwin Sands Shipwrecks*, pp. 53–59; Richard Larn, *Shipwrecks of Great Britain and Ireland*, David & Charles, Newton Abbot & London, 1981, pp. 65–70.
52 Quoted in *The Times,* 16 October 1982, p. 1; see also Stephen Martin, Journal, June 1702 to October 1708, British Library, Add MS. 47970, fol. 23v, entry for 27 November 1703; Stephen Martin, Logs, 1699–1707, British Library, Add MS. 47970, fols. 65r and 66r, log entries for 27 and 28 November 1703.
53 The *Stirling Castle* was a two-decker 3rd rate of 70 guns, built in Deptford dockyard in 1679, rebuilt at Chatham Dockyard in 1699. See David Lyon, *The Sailing Navy List*, Conway Maritime Press, London, 1993, pp. 20–21; J.J. Colledge, *Ships of the Royal Navy*, Greenhill Books, London, 2003, p. 309.
54 David Lyon, 'The Goodwins Wreck', *International Journal of Nautical Archaeology*, 9 (1980), pp. 339–342. Another cross-staff fragment found more recently is described in B.S. Smith, 'A Cross-Staff from the Wreck of HMS *Stirling Castle* (1703), Goodwin Sands, UK, and the Link with the Last Voyage of Sir Cloudesley Shovell in 1707', *International Journal of Nautical Archaeology*, 39 (2010), pp. 172–181.
55 Peter Ifland, *Taking the Stars*, pp. 7–9.
56 Willem Mörzer Bruyns, *The Cross-Staff: History and Development of a Navigational Instrument*, Zutphen Walburg, 1994, pp. 31–34, 56.
57 Bruce E. Babcock, 'Some Notes on the History and Use of Gunter's Scale', *Journal of the Oughtred Society*, 3.2 (1994), 14–20; Otto van Poelje, 'Gunter Rules in Navigation', *Journal of the Oughtred Society*, 13.1 (2004), pp. 11–22.
58 The *Britannia* was originally the *Prince of Wales*, but was renamed when she replaced a previous *Britannia* at Dartmouth. The first *Britannia* had taken over from HMS *Illustrious*, the first cadet training ship there.
59 *Life on Board H.M.S. "Britannia"*, J. Gieve & Sons Ltd, Portsmouth, c.1900; E.J. Felix, *How to Enter the Royal Navy*, Simpkin, Marshall, Hamilton, Kent & Co, London, 1899; E.P. Statham, *The Story of the "Britannia" the Training Ship for Naval Cadets*, Cassell & Co, London, 1904, Appendix V.
60 H.D. Turner, *The Cradle of the Navy. The Story of the Royal Hospital School at Greenwich and at Holbrook, 1694–1988*, William Sessions Ltd, York, 1990.
61 Quoted in 'The Royal Naval College, Greenwich', *Nature*, 15 (1877), p. 512.
62 Susannah Fisher, 'The Origins of the Station Pointer', *International Hydrographic Review*, 68 (1991), pp. 119–126.
63 Gloria Clifton, *Directory of British Scientific Instrument Makers 1550–1851*, Zwemmer/National Maritime Museum, London, p. 221.
64 The Royal Naval School building is now part of Goldsmiths, University of London.
65 Robert Neal Rudmose Brown, 'Nares, Sir George Strong (*bap.* 1831, *d.* 1915)', rev. Margaret Deacon, *Oxford Dictionary of National Biography*, Oxford University Press, Oxford, 2004, www.oxforddnb.com/view/article/35185, accessed 10 June 2006; Margaret Deacon, *Scientists and the Sea*, 2nd ed., Ashgate, Aldershot and Brookfield, 1997, pp. 333–365.

66 George Nares, *Narrative of a Voyage to the Polar Sea During 1875–6 in H.M. Ships 'Alert' and 'Discovery'*, Sampson, Low, Marston, Searle & Rivington, London, 1878, pp. xvi–xvii.

67 *The Thames Conservancy 1857–1957*, London, 1957. The Thames Conservancy was set up in 1857 to look after the river between Staines and Yantlet Creek. The PLA took over responsibility for the tidal Thames below Teddington Lock in 1909.

68 See collections.rmg.co.uk.

Chapter 11

1 John Bold, *Greenwich: An Architectural History of the Royal Hospital for Seamen and the Queen's House* (New Haven: Yale University Press, 2000).

2 Thomas Hinde, *An Illustrated History of the University of Greenwich* (London: University of Greenwich, 1996), pp. 26 and 78–79.

3 Keith Rennolls, personal communication; Keith Rennolls and Mingliang Wang, 'Enhancement of image-to image co-registration accuracy using spectral matching methods', in M. Caetano and M. Painho (eds.), *Proceedings of the 7th International Symposium on Spatial Accuracy Assessment in Natural and Environmental Sciences* (Lisbon: Instituto Geográfico Português, 2006), pp. 296–305.

4 University of Greenwich, no date, 'The Fire Safety Engineering Group' [http://cnmpa.gre.ac.uk/group_fseg.html, accessed 5 August 2018]; *The Guardian*, 2014, 'The Guardian University Awards: Winner: the University of Greenwich' [www.theguardian.com/higher-education-network/2014/feb/18/winner-university-of-greenwich-research-impact, accessed 5 August 2017].

5 University of Greenwich, 2016, 'Top Prize for PhD Student's Evacuation Research' [www.gre.ac.uk/ach/about/news/ach/a3695-top-prize-for-phd-students-evacuation-research, accessed 5 August 2018].

6 Nick Jackson, 2007, 'How Greenwich University is helping to rebuild the Cutty Sark', *The Independent* [www.independent.co.uk/news/education/higher/how-greenwich-university-is-helping-to-rebuild-the-cutty-sark-432501.html, accessed 14 August 2018].

7 Channel 4, 'Food unwrapped', Series 12 episode 1 (2017), [www.channel4.com/programmes/food-unwrapped/episode-guide/, accessed 5 August 2018]; The University of Greenwich, 2017, 'Walking on custard', *Youtube* [www.youtube.com/watch?v=FMfc2t5wYzc, accessed 5 August 2018].

8 IMA, 2018, 'Tomorrow's Mathematicians Today 2019' [https://sites.google.com/site/imatmt2019, accessed 24 December 2018].

9 The University of Greenwich, no date, 'The Greenwich Maths Centre' [www.gre.ac.uk/mathscentre, accessed 5 August 2018].

10 Greenwich Maths Time – the 2017 IMA Festival of Mathematics and its Applications [www.gre.ac.uk/mathsfestival, accessed 5 August 2018]. Tony Mann, 2017, 'Greenwich Maths Time Festival', *Mathematics Today*, Vol 53 No 5, pp. 194–195.

11 Noel-Ann Bradshaw, Mark McCartney and Tony Mann, 2010, 'Bite-sized History of Mathematics Resources for use in the teaching of mathematics', [http://scm.ulster.ac.uk/~e10028898/mark/hom.htm, accessed 5 August 2018].

12 @STEMinStyle [https://twitter.com/STEMinStyle, accessed 24 December 2018].

13 Higher Education; Science, Technology, Engineering and Mathematics: the National HE STEM Programme was an initiative to encourage more students to study in these disciplines.
14 N. Bradshaw and P. Rowlett (eds.), 2012. 'Maths Arcade: stretching and supporting mathematical thinking', MSOR Network. [www.mathcentre.ac.uk/resources/uploaded/mathsarcade.pdf, accessed 5 August 2018]; Noel-Ann Bradshaw, 2017, 'The Maths Arcade: A tool for supporting and stretching mathematics undergraduates', in L.N. Wood and Y.A. Breyer (eds.), *Success in Higher Education: Transitions to, within and from University* (Singapore: Springer).
15 *MSOR Connections* [https://journals.gre.ac.uk/index.php/msor/index, accessed 5 August 2018].
16 **sigma** network for excellence in mathematics and statistics support [www.sigma-network.ac.uk, accessed 5 August 2018].
17 The Undergraduate Ambassadors Scheme [https://uas.ac.uk/, accessed 5 August 2018].

Chapter 12

1 Sir James Thornhill, *An Explanation of the Painting in the Royal Hospital at Greenwich*, London, 1730?; facsimile published by the Old Royal Naval College in 2017.
2 Ann Stewart Balakier and James J. Balakier, *The Spatial Infinite at Greenwich in Works by Christopher Wren, James Thornhill, and James Thomson* (Lewiston: Edwin Mellen, 1995).
3 For further information see the article by Alex Fedorec and Tony Mann in the 'Events and Sightings' section of the *IEEE Annals of the History of Computing*, Vol. 26, No. 1, 2004.
4 This information comes from John Bold, *Greenwich: An Architectural History of the Royal Hospital for Seamen and the Queen's House* (New Haven: Yale, 2000).
5 John Cooke and John Maule, *An Historical Account of the Royal Hospital for Seamen at Greenwich* (London, 1789), quoted by Bold.

Index

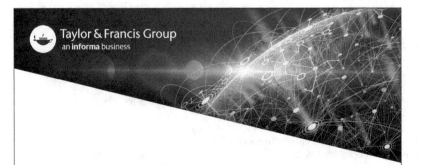